Environmental
Vacations

Environmental Vacations

Volunteer Projects to Save the Planet

Stephanie Ocko

John Muir Publications
Santa Fe, New Mexico

John Muir Publications, P.O. Box 613, Santa Fe, NM 87504
© 1990 by Stephanie Ocko
Cover © 1990 by John Muir Publications
All rights reserved. Published 1990
Printed in the United States of America

First edition. Second printing March 1991

Library of Congress Cataloging-in-Publication Data

Ocko, Stephanie.
    Environmental vacations : volunteer projects to save the planet /
Stephanie Ocko. — 1st ed.
        p. cm.
    Includes index.
    ISBN 0-945465-78-5
    1. Science projects. 2. Vacations. 3. Volunteer workers in
science. 4. Environmental education—Activity programs. I.Title.
Q182.3.O24   1990
363.7'0525—dc20   90-13245
                                                              CIP

Cover design: Sally Blakemore
Designer: Harry Wich
Cover graphics: Jim Wood and Holly Wood
Cover photographs: Dwight Sieggreen
Typeface: Garamond Book
Typesetter: Copygraphics Inc.
Printer: Banta Company
The text pages of this book were printed on recycled paper.

Distributed to the book trade by
W. W. Norton & Company, Inc.
New York, New York

Contents

Anthropology expedition in Tanzania. Photo: UREP

Preface

The purpose of this book is to help the prospective paying volunteer choose an environmental vacation on which he or she can work to make a difference in some part of the world. Saving the planet begins with understanding a small part of it; environmental vacations give volunteers not only that kind of understanding but also a chance to help change something that is wrong.

A word of caution about the word "vacation," however; these are not vacations in the sense that you loll about poolside thinking about what you will have for dinner. They are work projects that you do during your vacation time, but the emphasis is on *work*; healthy adults have been known to sleep for twenty-four hours after two weeks on a "vacation" like this. Some trips are more labor-intensive than others.

The environmental vacations discussed in this book are volunteer science projects that involve helping scientists in their fieldwork, on land and at sea, and others that involve helping people in need. A unique project in Arizona invites volunteers to help build a sustainable city. This book includes experiences of scientists and volunteers and outlines several typical science projects to give the reader a sense of what it is like to be there. The opportunities mentioned here are not intended to be a complete list, nor do they comprise a catalog. They are offered to help direct you to the organization that might come closer to providing you with the type of vacation you want.

Because these expeditions often involve travel to rural areas in less-developed countries for two weeks or more, I include a medical advice section based on the experiences of scientists and volunteers, with references to detailed information about traveling in high-risk areas. The information is offered in good faith and does not pretend to be complete. Always consult your own doctor before and after traveling to a high-risk area.

Thanks to Delores Schmitz, Christine Gabriel, Mike Hofhines, Jim Webber, Pansy Collins, Dr. William Robinson, Kay Wallis, Katy Eagan, Saurabh Misra, Thomas Banks, Ann Ballin, Jean Colvin, Mark Richman, Dr. Bruce Bradley, Dr. Carl Hopkins, Dr. Richard MacNeish, Dr. Irwin Ting, Eddie Bilezikian, Vicky Zack, Dr. Katherine Wynne-Edwards, Helen John-Castelli, Claudia Cline, Andrew Forster, Dr. Grady Webster, Robert Schilling, Dr. Kevin Padian, Howard Kaplan, Blue Magruder, Dr. Dan Jaffe, Adam Frankel, Sister Juliana Lucey, s.n.j.m., Evelyn Angeletti, Dr. Rick Chesher, Dr. Bill Melson, Leslie and Steve Bailey, Phyllis Hoo, Scott Davis, David Mowat, Dr. Yosihiko Sinoto, Cady Goldfield, Steve Ocko, Joan Thuebel, Gary Alpert, Rob Pudim, Sue Wolff, Molly O'Malley, Dwight Sieggreen, Jane Libby, the staff at Crow Canyon, the staff at Habitat for Humanity, the staff at Amigos de las Americas, Susan Urstadt, Dick Woods, and Peter Ocko.

Introduction

Tourism has probably been around since Neanderthal man crossed a river to track down mammoths. Over the years, it has taken on various shapes and hues, depending on the energy of the traveler and what there was out there to be visited. In ancient Greece, for example, Herodotus spent several years on the road, filing reports on the unusual things he saw and the people with whom he talked. In the Middle Ages, European tourists traveled on pilgrimages to holy sites, as far away as Jerusalem. A couple of centuries later, when traders made serious indents in the North American continent, the open routes of trade and settlement gave a whole new meaning to travel.

About a century ago in this country, travel had a real purpose. While immigrants were arriving to settle permanently, rich Americans were making the Grand Tour of Europe with an intent to own a piece of it. Wealthy people met each other in Paris and Rome where they rented houses for whole seasons, hobnobbed with the demimonde, and tried and often succeeded in marrying their daughters to titled but penniless Europeans. World wars altered that pattern, but Americans still perceived Europe as the locus of civili-

1

zation and savoir-faire. The obligatory tour after college graduation in the 1950s was considered a short-course finishing school: there the educated twenty-year-old was introduced to serious art in serious museums, hotels with bidets in the room, and strangers who pinched young ladies' bottoms. The 1950s and 1960s were also the time of the If-it's-Tuesday-it-must-be-Belgium tour in which all of Europe could be ingested in a few days by a tourist who was too tired to remember his or her manners. Tensions developed, and the Ugly American appeared.

In the 1980s, designer travel was born. Those who had done the usual routes were desperately seeking alternative travel. The 1960s college dropout who had toured Nepal wanted something more than tours offered by private organizations or universities which carried passengers, often by ship, to various places, usually with one or two experts to comment on the geology or the ancient history of the area. Other tour organizers ventured into places never visited *deliberately* before, where remnants of lost ships might litter the shore. The poles became popular, as did the deepest parts of the Amazon jungle. As interest in physical fitness grew, and people were amazed to discover they could push their bodies into new levels of endurance, adventure travel, usually featuring white-water rafting or mountaineering trips, offered guides and guidance climbing dangerous slopes or shooting wild rapids. These are so popular they are now graded according to difficulty, so an aficionado can graduate to the highest level of endurance and test himself against the best, while still remaining technically a tourist.

Tour design has probably never been as fluid as it is today. For various fees, you can balloon over the Andes, stay in an underwater hotel, meditate with monks. It's not that the search for the unusual is new; the difference is that in the past, the unusual created tours, not the other way around. Niagara Falls, the Taj Mahal, the Great Wall of China have all attracted their share of passive tourists, as have, unfortunately, American Indians, Bushmen, and Australian Aborigines. Sometimes it was a spectacle or a passing attraction. In Boston in the early 1800s, when a sea monster was said to have been sighted in the harbor, its prehistoric head rising above the surface of the water like a fun-house boat, anyone with a raft big enough to carry another person went into the tourist business for the month or so the monster stayed around. But now travelers demand interaction—with the environment, with people, or with animals.

Environmental awareness has put a spin on travel and has created a whole new vocabulary. As diet-conscious consumers created "light" products, so have environmentally aware consumers created a market for tours that somehow neither insult nor leave a trace on the environment, yet still satisfy. But in many cases, they are the same old tours done up in new buzzwords. *Environmentally sensitive, responsible, low-impact, alternative, softpath,* and *ecotourism* are some of the new labels that produce guilt-free travel. But the edges are blurry. Some tours engage in some kind of ecological "help," such as mapping a coral reef, which, if not done with professional guidance, can wind up being a piece of paper with some curvy lines on it. Others are what they have always been: luxury tours with an informed guide, the difference being that now the guide discusses the environment. In ecotourism, the emphasis is on encouraging host countries to preserve their environments or create game parks or reserves that will attract tourist dollars. The good intention is in place, to awaken awareness or to give something back to the area visited, but the industry is still forming.

As tourists clamber to jump on the eco-bandwagon, anthropologists and sociologists struggle to keep an eye on the rapid changes this is causing. In fact, it seems that all tourism leaves a trace of some kind: there's no such thing as no-impact tourism, and low-impact tourism might be a misnomer, too. The problems are subtle, in some cases. The subculture of hotel workers, for example, is the subject of a lot of anthropological study: for the most part, they speak a few words of several languages, communicate well in none, and exercise a lot of power in situations where tourists are at their mercy. Very often, tourism can alter the rank and file of a native group and elevate some over those who are actually their superiors according to their society's organization. While tourism dollars can introduce a cash economy, sometimes it does so at the risk of upsetting native sacred rites, as for example, when Native Americans are asked to perform sacred dances for a tourist camera.

The biggest change in travel is in the trend toward combining vacationtime with a sense of purpose. Learning vacations and work camps have been around for a long time. In the 1870s, the experiment begun at Chautauqua in upstate New York became immensely popular. There, whole families would stay in modest cottages and participate in lectures on philosophy, religion, or politics, or musical concerts or dance, and share ideas like ancient Greeks in sylvan surroundings. Work camps were designed to introduce students to the importance of hands-on labor—building structures, making pottery to use, working in the fields. People-to-people programs began after World War II, with citizen and student exchange programs in which it was possible to live with a family in a foreign country and thereby share ideas and foster understanding of another culture.

All of these still exist, but in the last ten years, they have taken on a new shape and importance. They all come under the rubric of "environmental vacations," because they foster sensitivity to other cultures and environments, create new understandings, and give the participant a sense of connectedness that global communication implies. It's one thing to watch television and be aware of world problems, or to know obscure facts about obscure people; it's better to be able to experience and meet them firsthand.

Volunteer Science

Volunteer science began in the 1970s when government funding for scientific fieldwork was cut back and scientists needed short-term labor gangs, as well as money. Archaeologists have often hired local labor to help with excavations, and scientists usually take graduate and undergraduate students into the field with them. Few precedents existed for interested amateur volunteers who wanted to come along; and there were no precedents for paying volunteers, those who were willing to pay for the privilege of giving their labor to a scientific project.

The concept of the paying volunteer on science projects was helped in part by media attention to such science superstars as the astronauts who walked on the moon, underwater explorer and environmentalist Jacques Cousteau, and the fabled Indiana Jones. The message was clear: science was sexy and glamorous, and it looked relatively easy.

It remained only to iron out the difficulties of moving a group of interested and diverse volunteers into a remote, often challenging location, making sure that the volunteer had a sense of learning something and that the scientist could actually get and use the data he got from the volunteers.

What has happened is scientists find that, enmeshed in the nitty-gritty of their labor at the ends of the world, they welcome people whom they would normally never meet, who arrive happy to help, bringing whatever is unique to their backgrounds. Paying science volunteers are thrilled with the opportunity to be involved in the frontline of science fieldwork, actually being able to touch and work with a wild animal in its own habitat, to hold an artifact buried for several millennia, or to talk with indigenous people in their own homes.

People-to-People Vacations

At the same time, the global people-to-people tour, formerly reserved for politicians on junkets and journalists on assignment, is growing in popularity. Here, organizers create travel for interested groups or individuals to politically sensitive areas such as Latin America, the Middle East, or South Africa and arrange interviews with local officials and spokesmen. Like volunteer science expeditions, these tours minimize comforts and maximize interaction. Their main thrust is to increase understanding with firsthand experience.

You can even organize your own deep travel: Co-Op America, with Schilling Travel, will arrange for you or your group to study a foreign language in a host country or will organize a special-interest tour to wherever you want to go.

Condensed "Peace Corps" Vacations

The Peace Corps, begun in the 1960s, demands a commitment of at least a year or more. Because of this and because it involves surrendering your own life-style for that of your host country, it has gradually declined in popularity. Although many people have the spirit, few have the time.

But now Peace Corps-type opportunities that take as little as a week are available in this country or abroad. Habitat for Humanity, for example, allows a person to help build houses for those who are without adequate housing here and in several countries around the world. Many student groups sponsor helping vacations over the summer or spring break. Some volunteer science projects are actively involved in helping a host country.

Adventure Travel

Strictly speaking, adventure travel involves a physical challenge. A hundred years ago, adventure travelers were probably big-game hunters or romantics trying to find a purpose in life by spending a night on top of Cheops' pyramid or swimming the Bosporos in Turkey. Adventure travelers push away boundaries and like to ride on the edge, accepting either the physical endurance involved in climbing impossible mountains or negotiating a raft over a waterfall or traveling in little-visited countries in difficult climates.

But as more volunteer science vacations take place in exotic fields, "adventure" is ceasing to have its exclusive meaning. One science volunteer recalled being on a botany expedition to the Sepik River and looking up from his dugout canoe to see a ship used by a large adventure company to penetrate the depths of New Guinea. "There we were chugging along with our little outboard motor, and the adventurers were up there sipping their gin and tonics. They slept in air-conditioned berths; we stayed in the actual village," he said.

Ecotourism

Distinctions in travel under the umbrella term "ecotourism" are still forming. The intention of ecotourism is to include host countries in any profit taking from tourism with the understanding that the money will be used to create beneficial environments for wildlife or to protect indigenous culture. In some cases, vacationers engage in what could be called "light volunteer science," such as banding birds or catching fish for aquariums. In most cases, tours are accompanied by natural history experts whose purpose it is to inform the interested tourist and to inspire a continued concern for the place visited.

On paper, the concept of volunteers paying to participate in environmental projects seems like the biggest hoodwink since Tom Sawyer got out of whitewashing his fence. But it works. Giving not only money but labor to a project, whether inoculating people in Africa or testing polluted waters in South America, allows a person to connect intimately with what is important in the world. The one or two weeks or the summer that it takes permits the experience to be incorporated into a person's life without dominating it. Most of the time, paying volunteers don't waste a precious minute. The sense that vacationtime is valuable for others as well as oneself is in itself rewarding.

Street scene on the island of Tonga. Photo: UREP

Chapter One

Environmental Vacations: Club Med with Horn-Rims?

"We're a nation of explorers. It's a fundamental thing to want to go someplace and feel and touch and learn."
—Astronaut Michael Collins on the twentieth anniversary of the first landing on the moon, July 19, 1989

"The scientific world will actually recognize the data I collected. I've never done anything that really made a difference before."
—First-time volunteer

In the tiny kingdom of Tonga, in the South Pacific, Rick Chesher does everything with a sense of urgency. A marine biologist for twenty years in the South Pacific, he has lived on a boat for that long and knows that if you don't do something immediately it can burgeon into a full-blown problem. It's very early in the morning, and he's packing gear into the Zodiac which members of an Earthwatch team hand to him, talking all the while about what happened to

11

the sailing yachts in this northern Tonga harbor during a hurricane six years ago. Then, after the team steps gingerly into the Zodiac and sits along the edge, he yells, "Watch your nose!" and pulls the outboard into roaring gear, speeding across the harbor to his 44-foot cutter sloop, the *Moira*.

This is Chesher's third summer working with science volunteers. For most of the ten team members, it's their first time in the South Pacific, their first time working with a scientist, their first time as adults living with complete strangers, their first time living among hundreds of pigs, their first time living without hot water, their first time climbing from the trembling Zodiac up the ladder to the deck of the *Moira*, and certainly the first time they have ever measured giant clams underwater.

Although the volunteers have come for reasons as diverse as wanting a "scientific experience" to needing to get away and think, Chesher is very clear about his objectives. Years and years of overfishing in Tonga have made the giant clams —those with huge scallop-edged shells—nearly extinct. The clams are a small part of a much larger problem: the whole coral reef the *Moira* sails over is being destroyed by overzealous fishing, he says. Chesher wants to collect what's left of the adult giant clams and put them in breeding circles that he will instruct the Tongans to care for. It is his hope that in three or four years, a beautiful species can be saved.

He tells the team this. But to them, Tonga seems like paradise. The sea is an uninterrupted stretch of sapphire. Up ahead some tiny white sand islands rise up out of an emerald pool. The volunteers uncap the water bottles they have brought, take a swig, and as the *Moira* comes to anchor

Volunteer clam measurers relaxing on their boat as they return to home port on the island of Tonga. Photo: Ivan Strand

near one of the islands, they gather together snorkels and fins and shimmy into wet suits.

It's gorgeous, but this is not Club Med. It feels like the first day of school. No one says much.

Chesher explains the three types of clams, pointing out the minimal differences among their scalloped, shaggy exteriors and describing the colorful mantels of the living clams. He explains the use of the calipers and emphasizes that if they're dropped, the dropper goes after them. He asks snorkelers to hold their breath for five seconds, then for ten. "That's how long it takes to measure a clam," he says.

"What if we miss measuring a few?" asks a computer marketer.

"Do what you can."

"What about sharks?" asks a man who in another life is a corporate lawyer.

"Nah," says Chesher. "Sharks won't bother ya."

Underwater, the dead coral reef is as white as Yorick's skull. It goes on seemingly for miles. Waving around it are sea fans and anemones through which weave an endless commuter line of fish of all sizes, shapes, and colors. The snorkelers follow Chesher as he dives gracefully, with practically no bubbles escaping from his snorkel, clings to the coral, skillfully adjusts the calipers with his free hand, and measures the clam embedded in the coral. He rises like a seahorse to the surface to give the dimensions to a bobbing volunteer holding a clipboard.

At first it's hard to spot the tiny squiggly lines of the clams in the coral; then it's hard to keep from being pulled back to the surface; then it's hard to be absolutely sure the dimensions are accurate. But after a couple of hours, it becomes almost easy; within a couple of days, it's hard to imagine ever having done anything else.

By the end of the day, in bathing suits, sitting on the deck of the *Moira* as it heads back to port in the warm afternoon sun, the team members feel as if they have done something. It's not something they will be able to explain easily, even to close members of their families. Dangling upside down under the South Pacific and measuring clams in itself hardly seems like a noble cause, even for science. But as team members joke about one other's underwater styles and recount anecdotes from the day's challenges, using a whole new vocabulary, they have a sense of accomplishment in something that has absolutely nothing to do with their other lives. It's exhilarating.

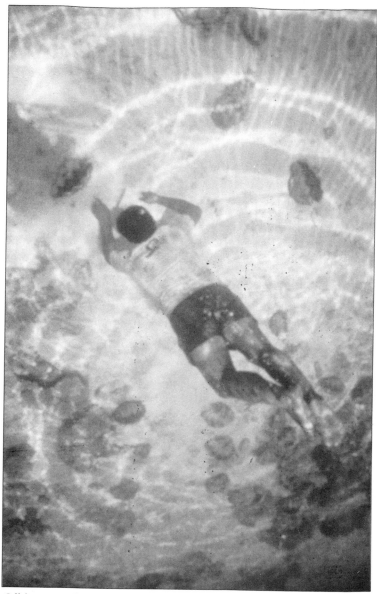

Off the coast of Tonga, a district commissioner dives to check the giant clams in their breeding circle on the ocean floor. Photo: Rick Chesher

At dinner that night at the Tongan hotel where the team stays, five men and five women, ranging in age from 25 to 60, now a social club of clam measurers, trade stories. Someone asks, "Does anyone know why the hell we were measuring clams?"

From the grab bag of new information, together they sort out that Chesher wants figures from clams in the sea to compare them with growth rates of clams in the breeding circles he has set up in another location.

But the breeding circles come tomorrow. At the moment, the task at hand is to maneuver through the night to the cottages and cold-water showers without annoying a gang of irascible pigs. The stars seem brighter here than elsewhere, and the thought of sleep after a day of working underwater is delicious.

At first glance, volunteer science vacations appear to have been designed for the gullible: for a hefty sum, you will be able to spend your vacation working your fingers to the bone for someone else. You will also get to bunk with strangers, possibly cook your own food, and use a latrine. The temperature might be 105°F all day everyday, and the guy in charge is no cruise director. Forget your designer swimsuit; you'll need industrial-strength shorts. The only stars you'll see will be in the heavens, not in a restaurant guidebook.

But listen to what these volunteers say two weeks later:

"That was the most thrilling experience of my life!"

"I'm euphoric. It's like *National Geographic*, but anyone can do it."

"I've been on luxury cruises, on tours to Europe and Asia, but this really took me away. It's foreign to anything else in my life."

"Hands-on personal experience at the cutting edge of discovery!"

Simply put, what happens is that volunteers help fund a scientist's fieldwork. In return, the scientist shares with them the agonies and glories involved in reaching his over-all scientific goal and enlists their help in part of his field-work. It usually involves saving a tiny part of the planet by correcting abuses, or redeeming an endangered species, or trying to better understand another culture. Middleman organizations bring volunteer and scientist together through catalogs that advertise projects that need help. The scientific projects are infinitely varied, ranging across all dis-ciplines under the sea and over the land: digging for arti-facts of the ancient Anasazi in the American Southwest, tracking hamsters in Siberia, listening for volcanoes in Yel-lowstone, counting monkeys in Liberia, bird-watching in China, helping to build a city of the future in the Arizona desert, and diving to a wreck off Belize. And the list changes every few months.

Pick any one; it is guaranteed to be exotic, tiring, chal-lenging, instructive, adventuresome, and transforming. Volunteers say they escape on an environmental expedition more completely than on any other vacation, probably because they are with people they have never met before doing something completely alien to what they normally do. And they come back, they say, knowing more about a lot of things, including themselves. Many have been moved to change their careers, go back to school, or restructure their personal lives after the experience.

Psychologists could analyze the components that make adult hands-on learning vacations so rewarding: the sense of play, for one. Just as kids can pretend they're streaming though Hyperspace to save Planet X from extinction, so can a fully grown adult in flippers and mask dive into the ocean with the approximate cool of Jacques Cousteau and help a scientist save a dying reef. For the volunteer, it is two weeks filled with the challenge of exploration, investigation, and imagination. Even the most hard-boiled volunteer can't miss the noble purpose when it kicks in, along with the realization that we have the power not only to destroy our environment but also to repair it.

Within the last twenty years, more than 30,000 people have gone into the field, most of them in the 1980s, when middleman organizations reported exponential leaps in the number of people seeking environmental vacations. After twenty years, volunteer science vacations have produced a whole new subculture of people with unusual skills that are applicable only in unique situations. Clam measurers,

A volunteer enjoying the company of Tonganese children. Photo: Ivan Strand

for example, are not in high demand everywhere. Some amateur archaeologists, middleman organizations report, spend every vacation on excavations and have become adept at what they do. One project scientist foresees the day when volunteers will become specialized and middleman organizations will draw up lists of those with expertise.

This type of vacation can become addictive. About one-third of science volunteers are returnees, those who have been on previous expeditions. Record holders—those who have done thirty trips or more—go on several projects a year. "Learning keeps a person plugged into life," said a volunteer.

Over the years, the vacations have become increasingly diverse. At Crow Canyon Archaeological Center in Colorado, administrators help accommodate parents with young children by offering special archaeology programs for kids while their parents work on separate projects. The family meets again in the dining hall at evening. Elderhostel, an organization that provides international, short-term learning courses for people over age 60, reports that some retirees hop into their RVs at the first sign of warm weather and travel around the country, stopping at campuses that offer Elderhostel programs. A UNICEF program coordinated with the University Research Expeditions Program (UREP) brings American and Ecuadorean teachers together at workshops in Ecuador.

Environmental vacations are an idea whose time has come. They are unlike any other vacation. They can be a challenge geographically, socially, and intellectually. This book will tell you about some of those unique challenges.

On California's Carrizo Plain, volunteers band and collect data on birds of prey. Photo: Foundation for Field Research

Chapter Two

The Financial Aspects: Where Does Your Money Go?

Paying volunteers are a unique breed. The very word *volunteer* exudes a kind of altruism. But *paying* gives it a dangerous commercial ring. What exactly are you paying for? Food, lodging, transportation at the site? A dream? The opportunity to be a scientist for a while?

You pay for all of these and more. Officially, your money goes to making you comfortable in the field and feeding you and providing any transportation other than getting to and from the site. A slice of it goes to the scientist for his research, often called a grant. The rest goes to the middleman organization for their administrative costs, the costs of attracting and recruiting volunteers through advertising and a published catalog, and the purchase of field equipment and vehicles.

Percentages that go directly to the scientist and his field project and the amount that goes into recruiting and advertising for all projects as well as for office costs vary wildly among the organizations. The Foundation for Field Research (FFR) feels comfortable sending 75 percent of the fee to the field project. The University Research Expeditions Program at the University of California sends about 65 percent to the project. International Research Expeditions (IRE) says it sends 100 percent to the project. According to the Earthwatch *President's Report for 1988-1989*, 41 percent of Earthwatch funds goes to "grants and awards."

Your dollars may or may not be used directly to fund your own particular expedition: most organizations disperse the money allotted for the fieldwork to help fund projects that are undersubscribed.

Comparatively, volunteer science is a good deal. Adventure vacations—trekking in Nepal, shooting the gorge in Zambia—cost more (for guides and equipment). But volunteer science vacations, from less than $600 to $3,000, will guarantee you an experience in another place unlike any you have ever had. If your altruism falters and the price looks steep, consider that there are probably perks: a sandy beach that is all yours for the weekend; an exciting city within driving distance on a weekend. Or, if you're in the middle of nowhere chasing the long-winged batfly, you might enjoy learning about the exotic fungi that thrive in the same place. Also, unlike the average vacationer, you will be much more sensitive to your vacation spot, both because you will be examining some scientific aspect peculiar to it and because you will meet, work, and socialize with the locals.

If you teach, you are eligible for a volunteer science scholarship; and if you are a student, you can apply for a scholarship and arrange to get academic credit, probably as independent study through your own university. Ask.

Returnees receive a discount: UREP gives 10 percent; Earthwatch gives 5 percent; FFR gives 10 percent on all subsequent expeditions after the third; IRE gives 5 percent if you recruit a new member. Crow Canyon sets aside two weeks for alumni, who are invited at a 10 percent discount.

The cost of your project, including travel to and from the site, is probably tax deductible, providing you don't stop en route. Any gifts or donations you make to the project or to the middleman organization (whether money or a tent) are also tax deductible. Check with your tax consultant for details. Some middleman organizations advise keeping a daily diary while on the project to record your time spent in the field. That way you will have something to prove you did what you said you did, and, as Oscar Wilde once said about his diary, "to have something racy to read on the train."

Volunteers in the Mojave Desert in California documenting the remains of prehistoric animals. Photo: Foundation for Field Research

The Rights and Duties of a Paying Volunteer

Remember that you are a paying volunteer. That gives you an odd privilege. "Paying volunteers," said one scientist, "are different from student help. You sort of have to tippy-toe around them, because they can leave at any time."

The concept of a paying volunteer seems to be uniquely American. "Why not work in a factory for two weeks?" asked a British archaeology student, amazed to see a group of paying volunteers on their knees at an excavation in Switzerland. In the Soviet Union, said one scientist, a volunteer is what the army uses when it wants a railroad bed dug.

American paying volunteers see themselves as buying a privilege to do something constructive with a scientist or in a needy community which they would not be able to do on their own. As one volunteer said, "I watched them on TV, the scientists going out into the field with helpers, and I said, 'How do you get to be a helper?' "

But the word "paying" implies that volunteers are buying something and should expect a certain amount of control over their experience in the field. Middleman organizations are very clear about how they define their own rights and duties, but no such document exists for volunteers. Although volunteers can recognize their obligations, where do they find their rights?

Most release forms that volunteers must sign before going into the field include a clause saying that if a volunteer's behavior is disruptive, he or she can be removed from the project. But, as volunteer Dick Woods, who is also a lawyer, noted, "It's never clear who makes the judgment call. And what exactly does it mean? That volunteers can't have

Working through the night, volunteers rescue the eggs of endangered sea turtles from the hands of black market poachers. Photo: Annie Gutierrez

an occasional beer? On my expedition," he continued, "we spent most nights at a local bar and some nights in a vacant cottage. But who would ever admit it? The biggest problem with trying to spell out the rights of the participants is that the people who prepare the forms aren't interested in protecting the participants. Earthwatch is the squirreliest; their form probably protects them best from a legal point of view, but they may not be as concerned about the volunteer."

A volunteer's primary recourse, if he feels cheated, is to complain to the middleman organization. Because their success depends heavily on what volunteers say privately about them, middleman organizations show concern for their volunteers.

But the organizations are sympathetic only up to a point. In the field, the strongest defense a volunteer has is an unwritten and unspoken right: if he doesn't like it, he can talk with his feet. But will he get his money back? Probably not.

After the volunteers are in the field, no middleman organization will refund any money. "We have already committed the money for research," says an Earthwatch spokesman, "and we make them aware of that before they go."

Further, there is no compensation for acts of God, such as storms that keep planes grounded and the volunteer from meeting the team. "It's my impression," said a volunteer whose arrival at the site was delayed by bad weather, forcing him to spend three nights in a hotel, "that once you're in the air you can kiss your money good-bye."

Fortunately, the percentage of volunteers who find they are unhappy in the field is small. Most volunteers cannot believe that they got as much value out of their vacation dollar as they did.

Chapter Three

Essential Questions: Will You Survive?

The environmental volunteer offers two very precious things: his labor and his money. These are no small investments. He also hurtles himself into an exotic locale to live with total strangers for two weeks. And he suspends his own realm of expertise to learn something completely different. It is important, therefore, that the volunteer knows how these programs are run and what he or she can expect to get out of them.

The following are answers to some of the questions that volunteers most frequently ask.

Can't I just join the scientist in the field instead of going through a middleman organization?
Not unless you personally know the scientist and are sure your presence is welcome. Middleman organizations act as liaison between scientist and volunteer. They make sure that the scientist provides adequate housing, food, and local transportation, advertises the assembly point, and settles any problems that might arise. Ideally, the middleman organization helps the volunteer feel confident and accepted.

From the catalog, how can I tell how difficult an expedition will be? Why aren't they graded?
Middleman organizations, unlike adventure companies, do not grade their expeditions because there are too many factors involved: not only terrain and climate but the quixotic nature of what is being studied. How do you determine the difficulty of chasing frogs? Or of measuring giant clams in ten feet of water? They both depend on breath control and hand-eye coordination, and no chart of standards exists for that. If you have any doubts about your particular trip, call the middleman organization, which will put you in touch with previous volunteers.

How important are such things as climate, altitude, and terrain?
Read the fine print. If the average daily temperature at your destination is 100° F, ask yourself honestly if you do well in heat. It takes about five days to adjust to extreme heat and humidity: will the project work allow you to take it easy for five days? Remember that 9,000 to 12,000 feet is the danger point for mountain sickness, especially if you are not used to altitude. If you are not used to heights, even an altitude of 5,000 feet can give you a headache and make you feel tired. If you are going to southern Europe, find out if the maddening *foehn* will be blowing. It drives locals bonkers, and if you are already feeling stress, it might not be the wind for you.

Mentally place yourself in the environment as it is described before you go. Talk to the staff at the middleman organization. Talk to people who have been there before. It could save you a lot of future discomfort.

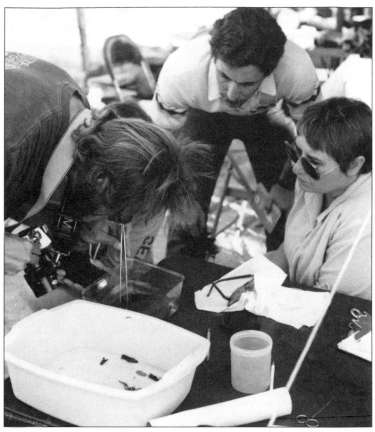

Nudibranchs being photographed in Baja California, Mexico. Photo: Foundation for Field Research

How much about the science do I have to know in advance? As much as you want. The project scientist will prepare a reading package to inform you what his big plans are and what segment of his research you will be working on. It will also include a bibliography. If it all seems like Greek before you go, don't worry. You will get training in the

29

field, where it will all come together. Volunteers are always able to understand what is going on, and the intellectual demands are reasonable. Volunteers are not expected to be whiz kids. "Volunteers are very talented people," one scientist said, "but you can't expect they'll arrive and be scientists for two weeks."

Scientists do stress that preparatory reading is highly desirable, although they recognize that few do it. You are also not expected to find the Missing Link or the Ark of the Covenant. "There is that aspect of discovery," said another project scientist, "but it's relatively low-level discovery."

Do I need a college degree?
Definitely not. Statistics indicate that more than 80 percent of science volunteers have some kind of college degree, but the main thing is not to be a slouch. You wouldn't consider being an environmental volunteer if you weren't interested in learning something. One no-nonsense project leader laid it on the line: "We have some very specific objectives we want to accomplish, and we're looking for team players who will attend the lectures, absorb what we're asking them to learn, and put in a full, honest day's work."

What if I have special skills?
One project scientist commented that his biggest problem in working with volunteers was not being able to elicit their hidden talents. If the situation doesn't come up, the volunteer is not likely to mention that he can fix a pickup, speak fluent Thai, cook for crowds of twenty or more, or iron out software glitches. But in extreme situations, every man or woman is a potential Superman. Don't forget that.

A volunteer banding the small, swallowlike American kestrel in California.
Photo: Foundation for Field Research

The middleman organization form you will fill out will ask for that kind of information, but if you feel you have some professional skill that is pertinent to the project, such as drawing, photography, or computer programming, by all means, point it out.

Who goes on these vacations?
No average volunteer environmentalist exists. One Earthwatch spokesman commented that people whose lives would never cross in any other context come together on expeditions. Ages range from 15 or less to any-age-still-able-to-do-it; the median age tends to fall in the early forties. Volunteers are from all geographic areas of the United States, including Hawaii and Alaska. Women hold a slight edge (60% at Earthwatch; 52% at FFR; 57% at UREP). Professionals make up the majority (52% at Earthwatch; 48% at UREP), with educators holding a close second, followed by students and retirees.

What happens if I miss the project leader at the assembly point?
Take the telephone number of the project site with you. Ask the middleman organization before you go what will happen if you don't make it there on time. You won't be the first volunteer to spend the night on a bench at the airport, but you will be missed. The project leader will have a list of all arriving volunteers. (See The Assembly Point, below.)

To what extent is the project scientist my boss? What if he tells me to do something I don't want to do?
No one is more acutely aware of the who's-the-boss problem than the project scientist. There he is blithely living in

the field, engrossed in solving his daily problems, when suddenly he is scoutmaster for ten or more healthy adults who give orders every day in their own lives. For the most part, the project scientist assigns tasks and responds when someone is unhappy. One scientist says his principal responsibility is to his volunteers when they are in the field. He assesses their abilities the first night at dinner and delegates tasks. "Everybody does something. Ideally, everyone does a little of everything," he says.

But not all project scientists are personnel managers. If you are confronted with an order you find distasteful, say so. You have paid to learn, and you are the master or mistress of your own destiny. But remember in *Mutiny on the Bounty* how long Fletcher Christian waited before he mutinied against the tyrannical Captain Bly. Sleep on it first. Still, be careful you don't come across like someone who should have gone to Club Med. Middleman organizations cannot screen out vacationers, and project scientists get very upset about volunteers who don't want to work.

Who arranges travel to the site?
You do. But call your middleman organization first because they've been there before. The Foundation for Field Research requires volunteers to book through a particular travel agency for block bookings; Earthwatch books some block seatings; and UREP, IRE, AFAR, and Elderhostel advise about the best connections to make.

Who's in charge of quality control in my field lodgings?
If you've heard that scientists can go for weeks without bathing and are happy sleeping on the ground, it doesn't

Loading up for a botanical project for the Museo Regional de Baja California Norte. Photo: Marc Robin

necessarily mean that they will subject you to the joys of roughing it. In most cases, the project scientist or his assistants will find a place that is affordable and habitable, whether it's a mud hut in Cameroon or a small hotel in France. You will probably sleep 2 or 3 to a room.

Middleman organizations set minimum standards: everyone gets a bed under a roof, and so on. And the scientist will probably want you to experience local color by keeping you out of the Hilton, if there is one. If you will be backpacking, the catalog will make that clear; some project leaders prefer experienced backpackers, which they will request.

What if I can't eat the food?
If you can't eat the food because you find it tasteless, tough, and unidentifiable, just wait until you're hungry. In fact, you're not likely to encounter too many foods that are really inedible anywhere in the world; and no country is without local or imported beer or wine. If you can't live without Snickers bars, bring them along. Actually, the combination of work and being outdoors produces a kind of spa effect. "These vacations are as good as fat farms," a volunteer commented. "I lost ten pounds without trying."

If you have a special diet, bring your necessary foods with you, if you can, or advise the expedition leader before you go.

What if I can't drink the water?
Where you will be staying will undoubtedly also have fire, and someone will boil undrinkable water for the team members. If you like, bring iodine tablets, or add 6 drops of

iodine per quart of water and let it stand for 30 minutes. Avoid ice cubes, and avoid brushing your teeth with tap water. The tiny microbes can cause diarrhea and give your periodontist a field day.

What will the bathroom be like?
Whether it's the hotel variety or an outdoor latrine that you build yourself, you *will* have a bathroom, although it might be a learning experience. "The bathroom facility might be a local bush," said one volunteer. "But when they dig a latrine, it's always private." The odds of sharing the bathroom facility with others are very high. If you use a latrine, you might want to limit your visits to daylight hours, and remember that spiders also like latrines because of the insects they can catch there. If you're sharing it with ten others, as one volunteer said, the operative word is speed. You do get used to it. Remember, you're doing it for science.

What about special services?
If you feel certain amenities are a necessity, for example, an occasional visit to a hairdresser, call the middleman organization to find out if it's a possibility.

What if I can't sleep?
After being outdoors all day, most volunteers can't wait to get to bed and have no trouble sleeping. Some volunteers advise against sleeping pills because they can make you very dim at work the next day, especially if you have to get up at dawn.

Housing for the field research trip members in Michoacán, Mexico. Photo: Walter Meyer

What about safety in the field?
No one will be careless about safety. One project leader said, "I worry about safety all the time." In work groups like these, the feeling of shared risk absorbs any unnecessary worry. On a particularly hazardous expedition, one volunteer said, "All is done that can be done to ensure personal safety, so there is no reason for me to be anxious."

Will I have free time to do other things?
If you are in the bush and completely away from civilization, you probably won't have a lot of "free" time. On most expeditions, you will definitely have some free time, either in the evening or on weekends, unless a scientist is really pressed to finish something.

Taking a break is essential on labor-intensive projects and the project leader usually plans some diversion. Some expeditions might require you to work odd hours, from dawn to noon, for example, to avoid extreme heat; or from sundown to sunup to track nocturnal turtles.

What can I do if I'm not happy with the expedition?
Breakdowns in communication are the cause of most problems on environmental trips, either volunteer vs. volunteer, or volunteer(s) vs. scientist. Project scientists come in all flavors, and many have never worked with anyone but graduate students or hired labor before. One archaeologist in the Middle East made his volunteers haul dirt all day, every day. Another researcher got angry because her volunteers could not move as fast as she could in their high-altitude locale.

Among volunteers, dissension can develop. An engaged couple ventured out on a project on a team composed of single men and women. They left struggling to hold their relationship together. One volunteer expressed his panic by refusing to work when he found the team different from what he expected.

Sometimes these vacations can be very stressful, with the combination of challenging climates, learning new material, eating differently, and sleeping in strange quarters. Carefully consider how to air your problems, and try communication and negotiation before anything else.

To help avoid things going wrong in the field, the Foundation for Field Research sends a field manager on each expedition whose sole function is to take care of day-to-day operations and to resolve differences if they occur. "Most scientists are ivory towerists, and they're incapable of dealing with daily problems," says FFR director Tom

Banks. Most middleman organizations require that the scientist bring an assistant or two.

What if I absolutely cannot do it?
Assess the following: (1) how much time you have left on your vacation; (2) how much money you will probably not get back; (3) what your insurance will cover if you leave and go somewhere else; (4) what the penalty is for your airline ticket change; and (5) if your reasons for quitting are reasons you will be able to live with when you're back home.

What are the odds of things going wrong in the field?
High. "In any field project, expect the unexpected," said one scientist. Scientists know too well that vehicles break down, it can rain for days on end, and pipes can burst on the day the local plumber leaves to visit his mother three villages away. But that's what makes a volunteer feel he or she is really on the frontline.

The operative word is flexibility. Don't go with preconceived ideas of what it will be like. Don't go expecting anything great. Don't go thinking the bugs will be on vacation. Don't go expecting to change the world. Murphy's Law will prevail. Ask any scientist. Scientists call it "chaos."

What if I get bored?
Say so. You are bringing intelligence to the project, and you are valuable as long as you are happy. The project scientist will assign you to another task. But you probably won't be alone if boredom strikes briefly at the end of the first week when "interest levels plummet after the initial enthusi-

asm," as one middleman organization director of field operations said. "But by the end of the second week, volunteers are broken in and are more intellectually and emotionally involved in the project."

How can I self-confidently operate in a strange, potentially hostile environment?
First of all, you will not be alone; everyone else on your team will feel as if he or she is in the same boat. The scientist and his assistants, or the field manager, probably know the area and can answer any questions you might have. Generally, the rule of thumb is that if anything—restaurants, food, animals, fish, the sky, the crowd in the town square—looks odd, there's probably something wrong. It won't take long to get a feeling for it. Trust your intuition.

Volunteer counting turtle eggs on the beach in Michoacán. Photo: Walter Meyer

What if my roommate is weird?
Take a walk. Individuals might not particularly like each other, but the group will probably bond, unless the chemistry is really skewed. Primates are sociable creatures, and as one middleman organization spokesman said, "If you take a small group of people, put them in a fairly tough situation, and let them solve their problems together, they form personal and emotional bonds." A volunteer said, "When I felt problems brewing, I just tuned out. Besides, there's always the science, if the social aspect gets cloudy."

One-week groups are more likely to hold sentimental reunions, as much as a year later, especially if most of the members are retired; but two-week groups generally disperse after the project, said a middleman organization spokesman.

Are volunteer science vacations just for singles?
Not at all, although the majority of volunteers travel alone. Still, many couples go together. Crow Canyon encourages parents and grandparents to bring children.

Will there be a social life?
The whole experience is social. Most groups never stop talking. Usually, expeditions have time for lengthy after-dinner conversations; some have campfires and guitars; others are near towns with ice cream parlors, bars, discos, bowling alleys, and pool halls. Runners can usually run wherever they are. After a hard day's work, volunteers look forward to relaxing together.

What do I need to do scuba diving?
Experience, guts, and caution. In addition to certification, you must show proof of recent dives. No middleman organization wants a scuba accident, so you will be closely monitored. UREP requires a test dive in a pool under the supervision of one of their representatives before you go. FFR field managers give a test dive in the field. "All the scuba documentation is meaningless; we must see it in action," said FFR director Tom Banks. Earthwatch checks record books, and it is up to the project scientist to give test dives in the field. One marine biologist says he runs his scuba dives like a military camp. "When they leave, they are better divers," he says.

However, you will probably only skim the surface of such complicated skills as marine archaeology, which requires hours of underwater work without supervision as well as a delicate precision that cannot be monitored. Most marine archaeologists are content to have volunteers retrieve artifacts or do survey or photography work on brief dives.

What if I'm not accepted?
You will be, unless you're wanted by the law in six states. For all their talk about screening applicants, middleman organizations place people where they want to be or diplomatically move them around to even out the distribution of volunteers per expedition. No paying guest is turned away. One middleman organization claims that all applicants are screened by the project scientist but admits that 90 percent of the time, the scientist is either too far-flung or too busy to bother with deciding whether to accept an

applicant. So the staff does it. "If someone is concerned about having to trek ten kilometers a day over rough terrain in the Greenland expedition, we look into that particular case and suggest they try another expedition," says a spokesman.

Can I sign up at the last minute?
Yes. All middleman organizations say that if space exists on a project, and you have all the necessary shots, a medical form, a visa (if required), and tickets, and if you can Federal Express the necessary requirements to the middleman organization, they will expedite your papers. Allow 7 to 10 days.

Volunteers watch a researcher at work at an Indian site in Illinois dating back approximately 9,000 years. Photo: Center for American Archaeology

Chapter Four

Middleman Organizations

Middleman organizations (MMOs) accept applications from scientists, some of whom they have actively recruited. On the application, the scientist is expected to justify his research as well as the need for extra labor in the field. Then he itemizes a budget (usually based on a team of ten) for lodging and food for volunteers as well as equipment that he will need in the field, from trucks to yardsticks. In most MMOs, the application is submitted to a review of four or six "peers"—other scholars in the same discipline—who comment on the value of the scientific work to be undertaken as well as on the scientist's apparent ability to lead a group of adult volunteers.

Then the MMO assesses the marketability of the project, advertises it in their catalog, promotes it in the appropriate markets, and hopes to attract the right number of volunteers, part of whose contributions will go to the science project (the rest goes to the MMO). It is more complex than a dating service, and probably less complex than real estate, but the process is the same: matching the right components so that everybody is happy.

Many MMOs do not communicate with each other. Like

Ford and General Motors, competition is fierce. Scientists will leave one and go to another or will start their own volunteer organization. Volunteers are free to go on expeditions sponsored by all of them. Some MMOs choose some volunteers to be field representatives, speaking on behalf of the MMO in their communities, but volunteers are fickle, and the market is volatile.

"Insects, reptiles, flowers," says Mark Richman, director of IRE. "How do you get people to go on these projects? I don't like to disappoint researchers, and it bothers me because it reduces everything to popularity." Large furry animals and dolphins are sure winners every time.

Public Affairs Director Blue Magruder at Earthwatch has become adept at the unusual art of attracting volunteers for what appear to be unpopular projects. One involved crushing termites to identify their chemical toxins. If that wasn't a turnoff, she said, it also was to take place in Guyana, famous site of the Kool-Aid suicides, around Christmas, when most volunteers stay home. "I didn't think we'd be able to find ten people out of 250 million who'd be willing to do it," she said. But Magruder found a film about termites that she circulated in potential markets, arranged a speaking engagement for the project leader, and managed to fill the project on time.

MMOs know that special interest groups are not reliable. Bird-watchers, for example, might not be attracted to a project watching just one species of bird. Magruder found that gulls have few takers among bird-watchers. IRE's Richman wondered where all the environmentalists were when he publicized his arid-zone research and oil-slick-in-Patagonia projects.

Whale watchers off the coast of Hawaii. Photo: Stephanie Ocko

MMOs have serviced a growing number of people within the last decade, but some see the number of their programs leveling off into a size that is comfortable. According to spokeswoman Ann Post, the Smithsonian Research Expeditions Program will be happy handling about twenty-five projects a year, five more than they have currently. Size and quality go together at UREP. Richman foresees a time when MMOs will be out of date, and volunteers will be able to go directly to the scientist or organization that advertises its need for volunteers, in publications like environmental magazines, for example.

Earthwatch

"We're the granddaddy of them all," says Magruder. Although Earthwatch might not have originated the idea of the paying volunteer, today it is the largest middleman

47

UREP Director Jean Colvin with Barabaig women in Tanzania. Photo: UREP
photo file

organization. From its informal beginning in 1971, when a couple of scientists took volunteers on expeditions "studying rocks and stars, things that volunteers couldn't hurt," the organization has grown tremendously and can now boast of having sent 23,000 volunteers, called the Earth-Corps, into the field. In fact, volunteer science and Earthwatch are almost synonymous.

The Earthwatch headquarters occupy the better part of a wing of a former elementary school, its Gothic brick exterior overlooking a playground in Watertown, a suburb of Boston. It employs a staff of about fifty who manage the 100-odd projects they fund every year, and deal with the problems of safely transporting about 3,000 people into the field. Its president, Brian Rosborough, has been with Earthwatch since its inception and foresees its future importance. In his *President's Report 1988-1989*, he said, "The velocity and impact of global change need to be assessed quickly, if corporations, policymakers, educators, and scientists are going to have time to plan for the sustainable use of natural and cultural resources. That is our calling."

A few years ago, Earthwatch established offices in Great Britain, the Soviet Union, and Australia, where it provides volunteers for projects with British, Russian, and Australian scientists. Once a year the organization sponsors a conference for their scientists and volunteers in Boston.

Earthwatch offers scholarship programs for students and teachers. It gives a discount for recruiting a new member; it offers a 5 percent discount to repeaters and families of three or more. An annual membership fee of $25 brings a full-color magazine/catalog six times a year. Earthwatch gives a full refund if you cancel more than 90 days before the project and no refund if you cancel after 45 days before the project; but it will transfer 50 percent of your money to another project less than 45 days before it begins.

University Research Expeditions Program (UREP)

Jean Colvin exudes the crisp efficiency of a dedicated schoolteacher: she is neat, quick, attentive, and responsive. Then she mentions matter-of-factly that her first experience in West Africa was when she hitchhiked across the Sahara Desert. The architect of her own explorations, Colvin now circumscribes the boundaries of the University of California Research Expeditions Program. In its fourteen years of existence, UREP has offered high-quality scientific projects for paying volunteers led by scientists drawn from the University of California. "We're getting more into social and environmental issues," says Colvin. "It's what I perceive as a need, not a trend."

Getting host countries involved with UREP's scientific work is one of Colvin's main objectives. A new joint program, funded by the National Science Foundation in cooperation with UNICEF, will allow environmental teachers to work with nationals in Brazil and Ecuador. In a summer program in Ecuador, teachers will remain an extra week after the scientific project to work with local educators. The intent is to share curriculum ideas on science projects, using U.S. models and incorporating Ecuadoran needs. UNICEF will provide local contacts and contribute to what Colvin calls "a global perspective" that she sees as being essential, especially for teachers. This she hopes to

Members of a botanical expedition in East Africa. Photo: Frank Cannon

do through mutual sharing of ideas and experiences. UREP has a tradition of courageous innovation. In 1986, it launched a program called SHARE (Science Serving Humanity and Research for the Environment). On these projects, UC scientists and paying volunteers were able to bring their expertise to bear on the real problems of a country. And the problems were often crucial to the survival of the people. In Kenya, during the worst of the drought of the 1980s, volunteers helped desert nomads find water. In Brazil, where children were sometimes the last to eat during famine, scientists examined exactly what kind of social change could make a dramatic difference in the lives of the children. As a result, the government of Brazil instituted a school lunch program.

One of the most innovative SHARE programs was a project

on which volunteers interviewed London health workers on attitudes toward AIDS. "I had some trouble getting that approved by the Faculty Advisory Committee," says Colvin. "They said it's too controversial; people who are interested in wildlife behavior aren't going to be interested in AIDS. But it turned out to be the most popular project that year, and it filled immediately. Some volunteers even took the attitudinal survey into their home communities and did it there." The project has since been repeated in Amsterdam and Rio de Janeiro. (See Anthropology, chap. 7).

Talk with Colvin easily goes back to Africa. Cultural change there is quick; one of her interests is to preserve the rapidly vanishing culture for museums. She has already done this successfully in Tanzania; now she wants to move into the deepest and most interesting heart of Africa, Mali, where mud houses look like huge wedding cakes and some dance masks are four feet high.

One of the causes of vanishing traditional art in Africa is art vendors' desire to provide what they think tourists want to buy. At an art market in Senegal, West Africa, Colvin says, she saw what looked like a female fertility figure from Ghana which was unusual in itself, but it was also decorated with red, white, and blue beads that she recognized from her days in East Africa. "I asked him about the beads, and he said he got the idea out of a book about Kenya," she said. "There may not be any traditional art left!"

Colvin's interest in East Africa, where she herself has led several expeditions, stems from her days at the University of Nairobi, when she worked on a project investigating the feasibility of domesticating game animals for meat production after 90 percent of the Masai cattle died in the drought of the 1970s.

The walls of her office on the Berkeley campus are covered with snapshots of some of the East Africans she has worked with over the years, including one whose university education she is sponsoring.

The UREP office employs about a dozen people in cubbyholes on a couple of floors in one of the older buildings on the Berkeley campus. They have about 20,000 people on their mailing list and have sent about 3,500 people into the field.

UREP gives grants to K-12 teachers who give courses in the natural and social sciences and offers a limited number of partial scholarships to students. A full refund, including the initial $200 deposit, will be given if the expedition of your choice is full or canceled. Up to 90 days before a project, you can transfer to another project or get your money

A volunteer on an AFAR dig at a New Mexican site examines an 8,000-year-old scraping tool he just uncovered. Photo: AFAR

back if you cancel. No refunds or transfers will be made after 90 days.

The Foundation for Field Research (FFR)

In the lush and rock-strewn hills east of San Diego at the top of a steep winding drive is a family ranch that doubles as the headquarters of the Foundation for Field Research. Big lazy dogs stretch out in the warm afternoon sun; a Land Rover waits to be repaired; the whole place feels more like Kenya than California.

In his cluttered office on the second floor of one of the outbuildings, Tom Banks explains that although officially he is the treasurer of the FFR Board of Directors, in fact, he is its only active and full-time member. The international board that meets annually to decide FFR policy and direction is composed of friends and colleagues who are scientists and museum directors. FFR's only other personnel consist of four full-time and two part-time employees who work in another outbuilding producing the quarterly newspaper-catalog, compiling the expedition briefings, and handling the details of the day-to-day organization of FFR's 30-odd science projects. Some are as close as Bakersfield, California; others, as far away as Liberia. A big barn on the ranch houses the camping, scuba, and automotive equipment, donated over the years, that is used on expeditions.

But low overhead is not the only thing that distinguishes FFR: more than other MMOs, it incorporates adventure into the overall science volunteer package, bringing the experience closer to the excitement that exploration generated a hundred years ago. Where possible, for example, the

A chimpanzee cracks nuts at a campsite in Liberia. Photo: Foundation for Field Research

teams use horses and mules for travel into remote regions. One volunteer recalled a botany expedition into the Baja Peninsula in Mexico over terrain too rugged for the team to carry tents. "So we put our sleeping bags on the ground," she said, "and slept under the stars. In the morning we

55

woke up to see palm trees swaying in the breeze and hummingbirds fluttering around in the branches.''

But Banks draws the distinction. ''Some people confuse us with adventure travel and think it's going to be fun. If you enjoy working with a scientist, it will be fun. We're not a travel organization; we're a brokerage for scientists, trying to bring them money and support.''

The linchpin that unites adventure with science is the field manager, a paid member of the team who often cooks, usually stays at the camp, and acts as ombudsman between volunteer and scientist, solving any problems, smoothing out any complaints, letting the scientist do science and the volunteer help the scientist. ''The field manager takes the burden of worrying about the details off the shoulders of the scientist,'' Banks said. ''The field manager goes out to the project site two to three days early, sets up a bank account [if it's in another country], hires drivers, gets a cook or finds out where to buy quality food, then runs the project so the researcher is free to concentrate on research.''

FFR's six field managers all have flexible jobs elsewhere: one is an attorney, another a beekeeper; others are a dog kennel operator, a carpenter, and a can maker for Continental Can Company.

A volunteer on the FFR plant-collecting expedition to Copper Canyon recalled, ''We took the train in from the coast, about an 8- to 10-hour ride. The field manager met us and we rode in the back of a pickup truck to the Indian village where we camped. The villagers invited us into their houses and showed us the inside of their cave. Every morning before we went into the field, we bathed in the

Volunteers carefully unearth Anasazi artifacts. Photo: Crow Canyon

river and played with the children. Then we went out with the researcher and gathered wildflowers. We were tired when we got back from our chores, so we were able to rest while the field manager cooked supper. Camp chores were voluntary. The field manager made breakfast and lunches to take into the field, and dinner. He was cool, calm, and collected. He also stayed with the camp, so we never had to worry about leaving things behind.''

Originally incorporated under the name Have Mule Will

Travel, the company changed its name in 1982 when Banks, an archaeologist, and two others wanted to dig a site on Cedros Island off Mexico's Baja Peninsula. In time, FFR began volunteer funding other scientists' fieldwork. In 1987, it began two long-term projects, one in Granada and the other in Liberia. Both countries, said Banks, are in need of all kinds of scientific help, from ethnography to zoology. In Grenada, Banks organized five different expeditions: an archaeological dig, an ethnographic study of the folklore and annual carnival, a plant collection expedition to study the rain forest, a study of extinct animals, and a sea-turtle project.

Alexander Peal, a member of FFR's board of directors and a friend, returned to his native Liberia to become head of Wildlife and Parks. But, with political unrest and subsequent social disintegration, Liberia has had little money or labor to spend on the preservation of animals. With the help of funding from FFR, Peal established a game park and hired all the local hunters as park rangers. "So far the project seems to be working," said Banks. "The rangers have a high status, are paid, and wear uniforms. Recently, they captured several hunters who came in illegally from the Côte d'Ivoire, and turned them over to the army." To date, volunteer projects have helped identify and count game as well as examine chimpanzee living areas for evidence of tool manufacture and use.

Logistics are difficult in Liberia; the price of gasoline is astronomical, and members of a splinter political faction stole the park vehicle and destroyed it. Beginning in 1991,*

*According to an FFR spokesman, if political conditions make it impossible to work in Liberia at that time, FFR projects scheduled to take place in Liberia in January-March 1991 will be operated instead in Sierra Leone.

FFR will sponsor ambitious archaeological and ethnographic surveys of the Liberian people, which Banks himself will lead and which are expected to last for several
years. (See Game Park, chap. 7.)

FFR has initiated innovative weekend projects for volunteers who want to be involved in a research project but lack
the time. In cases where sustained fieldwork is not fundamental to the integrity of the science—for example, trapping and banding huge redtailed hawks near Bakersfield,
California, to establish migration patterns, or diving off San
Diego to search for and investigate what appear to be early
anchor stones—the projects are successful. FFR also spon-

Volunteer gathering data on a hawk-banding expedition. Photo: Walter
Meyer

sors one-week projects and two-month expeditions, for example, at a Roman site in southern Italy, that attract serious students of archaeology for the summer.

FFR offers 10 percent discount for those who have been on three FFR projects; 50 percent discount for physicians who will act as the project doctor on certain expeditions; 25 percent discount for nurses; and teacher and student scholarships. FFR will transfer 100 percent of your contribution to another project if you cancel before 30 days; 80 percent after 30 days.

Age limits: 13-14 (or "mature") to any age. FFR requires that volunteers book with a certain travel company whose blocked space ensures the scientist's and field manager's low airfare.

Smithsonian Research Expeditions Program

The Smithsonian paying volunteer program draws on its considerable resources at the Smithsonian Institution, which includes scientists who are conducting fieldwork all over the world as well as the archival resources of the museum. Volunteers, for example, can help catalog some of their enormous collections, such as animal bones dug up in the course of archaeological digs. The program also uses volunteers to take part in activities on the Mall in Washington, where the Smithsonian is located, documenting such things as the annual Festival of American Folklife. Volunteer photographs do not have to be professional to be valuable on this project, which is designed "for an audience that is not yet born."

The baby of MMOs, begun in 1988 with five teams, the Smithsonian Research Expeditions Program will service twenty teams in 1991. Twenty-five teams a year is their ideal, according to spokeswoman Ann Post.

The Smithsonian offers scholarships for students and teachers, with special preference for members of minority groups. It will give a full refund if it is notified of cancellation 45 days before the project; after that, it will withhold the initial $300. No refunds will be made ten or fewer working days before the project. Full refund will be made if you cannot be accommodated on an expedition or if it is canceled.

(For an example of a Smithsonian project, see Geology, chap. 7.)

On an AFAR dig in a New Mexican cave, a volunteer brushes off a basket, later dated to 6,000 B.C. Photo: AFAR

International Research Expeditions (IRE)

"This is a small, shoestring operation. We have no university support, no flashy magazine. All of the volunteer contribution goes back to the field," says Mark Richman, director of IRE. After traveling in Central and South America in 1985, when he met scientists who needed extra help, he agreed to try to find them paying volunteers. Richman spends at least half his time running the White Horse Antiques Store in an upscale block of downtown Palo Alto where, on this afternoon, his description of South American bloodsucking bats is interrupted by a timid request from a white-haired lady for the price of a piece of stained glass.

IRE offers a banquet of about fifty projects, as varied as a search for prehistoric trade patterns in Thailand and the analysis of the worm that lives in the New England conch in Massachusetts. According to Richman, only about four or five of the projects can be depended on to be solidly booked. Many will have borderline subscriptions. His particular expertise is in helping the scientist construct a budget. "I have a good business sense of what volunteers are willing to pay. For example, on an archaeology project, volunteers won't pay more than $1,000 in the United States, but they'll pay more overseas."

Harder to assess is which projects will be popular. "The Anasazi projects in the Four Corners area are always popular, as are poisonous snakes in Australia. But a Costa Rican bird project that had eight volunteers last year had no takers this year. Anything to do with King Arthur goes, and the Greeks and Romans are always more attractive than the Middle Ages."

Richman is able to keep in close communication with his scientists and volunteers. "The best volunteers," he says, "are those who have been to the area and who now want to go back in depth. This is the way to do it."

Before he can explain about his bats and bandicoots project, the lady has decided to buy the stained glass, and Richman puts on his other hat as antique dealer.

IRE gives discounts of 5 percent for recruiting a new member and to families of three or more. If an applicant cancels at any time, the refund will be made only if a suitable replacement is found. IRE will not refund the initial deposit of $200 if the applicant decides not to go.

Elderhostel

Elderhostel began as a vision—of people of all ages being able to visit university campuses, live there in modest accommodations, and study a wide variety of subjects in short-term courses. The man who envisioned Elderhostel, Marty Knowlton, is the kind of person people remember. During the 1960s, he spent four years hiking alone across northern Europe, staying at youth hostels. While he was there, he got interested in the Folkeschules, which are Scandinavian heritage centers, where people can learn about the crafts and traditions of the area. Back in the United States at the University of New Hampshire, he recognized the space where he could make his vision come together: the dormitories at the university, which were empty during the summer. Why not have a cultural program there in the summer? And why not make it for older Americans since so many opportunities existed for younger Americans?

So, in 1975, Elderhostel was born. Nothing like it existed in this country. Since then, Elderhostel has grown into an organization that last year serviced between 20,000 and 30,000 people over age 60. Its mailing list of 195,000 reflects the number of Americans who choose to spend a great deal of their time in the classroom or doing hands-on work with scientists in the field at 1,600 campuses and research centers in the United States and in 40 other countries.

"Elderhostel attracts active, courageous people who are so interested in learning new things, they'll strike out in nonconventional areas," said spokeswoman Cady Goldfield. "It has a life of its own. People get so excited about it, they adapt their life-styles to it. I talked with a couple who was considering selling their home and traveling around the country to Elderhostel programs," said Goldfield. "In fact, they are a sort of counterculture."

Elderhostel offers one-week campus programs in the United States and Canada which cost about $250, including dormitory and meals. It also offers longer scientific projects around the world—from two to four weeks—that cost $1,600 to $4,500.

The Elderhostel office houses a small operation on the fourth floor of an older building—the kind with silver radiators that bang when the heat comes on—as close to the middle of downtown Boston as anyone could get. The Elderhostel catalog, in small newspaper format, weighs close to a pound and is loaded with what appear to be thousands of opportunities.

"As a kid, *National Geographic* was my flying carpet to adventure and knowledge. Now it's the Elderhostel cata-

Elderhostel program members sailing on the Sea of Galilee. Photo: Jim Harrison

log, and I don't just read—I do,'' wrote one Elderhosteler.

Meanwhile, Marty Knowlton is putting together something called Gatekeepers to the Future, designed to be a think tank for older adults. Stay tuned.

Andover Foundation for Archaeological Research (AFAR)

More like Indiana Jones than Indiana himself, Scotty MacNeish boxed his way out of the University of Chicago in 1949, having won the Golden Gloves Boxing Championship in 1938. For most of his career, he has been parrying with archaeological tradition to become a leader not only in American, or New World, archaeology but also in

Volunteers climbing up the slopes of a volcano during the Texas Caldera Project. Photo: Foundation for Field Research

archaeological techniques. The discoverer of the first evidence of domesticated corn in the Americas (two tiny corncobs at Teotihuacán, Mexico, dated to 7000 B.C.), he is currently working at a site in southern New Mexico that will yield the earliest dates of habitation in North America.

"We've brought four kinds of science together to talk about climatic change. We may be able to tell you the cycles of drought over the last 10,000 years, which can help us in the future. No one's ever had that kind of data," MacNeish said.

Using what he calls an interdisciplinary approach, Mac-

Neish does nothing halfway. He has brought together a pollen expert, a physicist, the use of a mass spectrometer at Harvard, and the wits of AFAR treasurer Bruno Marino. He is using carbon and nitrogen isotopic analysis of ancient bones to determine diet, and hydrogen isotopic analysis of soils to determine rainfall. By examining coprolites, an archaeological term for ancient human feces, experts at Texas A&M are able to tell ''exactly what the person ate the day before seven thousand years ago. And if he had amoebic dysentery.''

AFAR was founded by MacNeish and others in 1984, when MacNeish decided to form his own nonprofit middleman organization after his stint as an Earthwatch scientist. AFAR is different, because it sponsors five or six qualified graduate students a year who get a chance to spend a semester working side by side with MacNeish and to teach AFAR's paying volunteers.

In the field at the dig in southern New Mexico next door to the Fort Bliss Military Reservation, AFAR publishes *Notes from AFAR*, from which the following is excerpted:.

Good morning America. Here we are in beautiful Orogrande Base Camp, listening to the troops shouting, 'G for Gung Ho, E for Everyday...'' at 6:00 on a frosty morning. While they trundle out their tanks, we'll be eating breakfast and getting ready to go into the field. By 7:30, we're clambering into Gertie (yes, the same old fearless Gravel Gertie [a 10-year-old GMC truck-station wagon] who trounced up Box Canyon for 3 years).

The Setting

We're 4.4 miles from downtown Orogrande, with only the community of Newman (two bars, a derelict garage, and a couple of houses, just over the border into Texas) between us and El Paso.

Down the road a piece is the Orogrande National Forest, a lone cottonwood. At the heart of downtown, or perhaps uptown Orogrande (pop. 60-70), is the Oro Chico Café. Up the street is a barn full of antiques, the post office where Liz Rumsey tenders our mail, a frequently closed rock shop-cum-grocery store, and My Place, the local bar where the crew is becoming proficient in playing pool. Down the street is Kevin's Mechanical Fix-It Shop, with which Gertie is intimately acquainted. Off a side road is a volunteer fire station. Thanks to its chief, we have a typewriter ball to go with the typewriter loaned by NMSU.

The Excavations

We are working in two very dissimilar caves. Cueva Pintada, Painted Cave, gained its name from the paintings that have been found on the walls. Our untrained eyes found a little serpent and some spots of color that we couldn't identify; when Tom Todsen (former surveyor for AFAR) visited us, his trained eye spotted about a dozen more places that had been painted, including an animal form right above where we had been sweeping the shelf.

In an inner chamber filled with deep refuse, we found pounds and pounds of bones from a variety of fauna: rodents, birds, deer, elk, coyote. What we've found mostly in Pintada, however, is projectile points —over 120 in two places where hunters met over the centuries to swap tall tales and wait for game.

Even more exciting is our second cave, Pendejo, with refuse over a meter and a half deep so far (and possibly double that). Pendejo not only has excellent preservation but is unlooted. We expect it is going to produce corncobs, feces, sandals, baskets, wooden tools, and lots of other exciting things that will give us

A volunteer on an AFAR archaeological dig finds a 3,000-year-old arrowhead. Photo: AFAR

information about ancient diets, settlement patterns, and life-styles.

The Props

We're extremely comfortable in our quarters. All of us have beds with mattresses and bedding supplied by the army. We've turned two of our sixteen rooms into a kitchen-dining facility and eat breakfast and make our lunches there. The officers' recreation room has become our lab. Each of us has his or her own room; between each pair of rooms is a bathroom with a shower and lots of hot water. Really, we've lived in plusher-appearing places, but this one has more creature comforts.

Downtown in Orogrande at the Oro Chico Café we eat our main meal of the day. Al, its cook and owner, has provided us with a variety of dishes—Hog Rustler's Banquet, Nacho Soup (which he just invented), a variety of vegetarian dishes, baked fish, Gyompki, his wife's fabulous barbecue sauce, and other wonderful tasty and filling foods. Al is spoiling us dreadfully; in addition to plentiful food, he takes time to discuss the day's events, politics, and local personalities and relates old prospecting stories. (Everyone in Orogrande who can gets into the Jarales Mountains. There's gold in them thar hills.)

When AFAR is not in the Southwest, it sponsors digs in Belize to study Mayan trade networks and near the St. Lawrence River to uncover Iroquois activity. In 1991, Mac-Neish will enlist volunteers for an extensive underwater excavation off the coast of Belize where a flotilla of some thirty-one ships lies in the mud 30 feet down in crystal-clear blue water. These ships are remnants of a failed Guatemalan-Spanish attack in the 1770s on a band of wily Scottish and English pirates staked out in Belize, where they were wheeling and dealing in stolen goods. MacNeish has a theory that ''Belize'' is a corruption of Wallace, the name of one of the pirates.

Two weeks on an AFAR expedition, including everything except airfare, costs between $1,200 and $1,500. A $200 initial deposit is refundable if you decide not to go.

Crow Canyon Archaeological Center

The screen door squeaks at the Crow Canyon lodge, but if you get up really early and sit in one of the rush rocking chairs on the long porch as the sun slips over the La Plata Mountains, you might see a deer drink quietly at the pond. Most of the time, however, Crow Canyon buzzes with activity. ''Research can't be one person working in a vacuum,'' said Ricky Lightfoot, a staff archaeologist. ''Talking about things fuels the intellectual environment.''

Located in the high desert in southern Colorado, near the place where four states come together, Crow Canyon is an educational center set up to teach archaeological methods to interested volunteers of all ages. ''It's important to understand not the glamorous, the biggest, the best, but the excitement in little basic things,'' said Lightfoot, who excavated a site called Duckfoot (where a clay duckfoot turned up in one of the first excavations and its mate in one of the last, several years later). There Lightfoot was able to trace the lives of members of one family, who lived in the 800s, over the course of two generations.

Volunteers helping to uncover one of several hundred kivas found at Sand
Canyon. Photo: Crow Canyon Archaeological Center

Some of the excitement that archaeologists share with
volunteers is in trying to understand what motivated a very
mysterious group of people, called the Anasazi, who lived
there from about A.D. 400 to 1300. The architects of some
superb sandstone house clusters, one of which, Mesa Verde,
is the center of a national park, the Anasazi flourished and
then dramatically "disappeared." For a long time, archaeologists thought that severe drought might have killed them.
More recently, it is believed they simply left because of
drought and went south where they blended with the
Zuñis and the Hopis. (No one knows what they called
themselves: "Anasazi" means roughly "not our ancestors"

and was a name given by the Navajos who came after them.)

New evidence deepens the mystery of the Anasazi. At the sprawling site of Sand Canyon, archaeologist Bruce Bradley, with the help of paying volunteers, has uncovered a collection of rooms and numerous kivas, which were round sunken rooms probably used for religious ritual. Occupied for the short period of A.D. 1240-1300, Sand Canyon just has too many kivas for the number of people who appear to have lived there. Was it a ceremonial center?

"Something funny went on between 1250 and 1300," said Bradley, director of the site. "In 1300, the whole society changed dramatically." Not only did everyone migrate elsewhere but they left behind their portable religious ritual items. In an area perpetually plagued with drought, did they leave because of a particularly devastating shortage of water?

That was the theory two years ago. But recent studies by archaeologist Carla Van West, who analyzed the soil for each year in the thirteenth century, indicate that drought certainly was not a factor: the soil was healthy and plantable for every year.

"A lot of things that motivate human beings aren't simply that they can't make a living on the land," Bradley said. "People tend to hang in there; they adapt. It's the social and political things that are the motivating factors."

Bradley is considering the theory that some powerful political or religious movement or figure, like a revolution or a charismatic leader, drew people to new areas, which caused them to abandon perhaps not their whole religion but their ritual. "There was no more place for the old ways," he said.

The mystery of the Anasazi unfolds every year at Crow Canyon.

Paying volunteers spend a week immersed in Anasazi culture. They learn to identify artifacts and try to make and use them themselves to see what the culture actually "felt" like. They can spend time in a replica A.D. 800 pithouse, try to light a fire with a bowstring, grind corn with a large stone, and throw a stick using an atlatl. Bradley is a national champion atlatl thrower of sorts; he once actually used one to fell a bison as part of a graduate student project to replicate life in the Stone Age.

A new volunteer program will plant an Anasazi garden to chart the effects that humans have on it; and, in a few years, it will actually build an Anasazi village.

Everybody is welcome at Crow Canyon. Special programs for elementary schoolchildren, high school summer groups, families, grandparents and kids, teachers, and special interest groups, as well as returning "alumni" groups, are offered throughout the season from February to November. Volunteers stay upstairs at the lodge in dormitory-style rooms or in one of the 10 hogans, which are octagonal, split-log rooms, like those used by the Navajo, which sleep four. Robust meals are served in the dining hall. Talk never stops.

Center for American Archaeology

The Center for American Archaeology, in Kampsville, Illinois, is probably the oldest operating paying volunteer archaeological organization in the country. Under a program for adults organized in 1971 by Stuart Streuver, an archaeologist at Northwestern University (who went to

Colorado and organized Crow Canyon in 1985), Kampsville, as the center is known, has linked archaeologists with thousands of interested adults to teach digging and laboratory techniques.

Up until a couple of years ago, the main site was a place called Koster, which at one time was the largest excavation unit in the United States. Stretched out in a valley along a bluff, the excavation covered only a part of the area occupied by Indians known as "Early Archaic," who were hunter-fishermen who lived about 9,000 years ago.

After Koster, excavations moved to a place called Twin Ditch, also dated to 9,000 to 9,500 years ago. Here artifacts are in an excellent state of preservation because of dry conditions, which makes archaeologists rejoice. A huge house site waits to be dug. Volunteers are needed and welcome. An Indian culture exhibit at the Center displays many of Koster's artifacts.

Kampsville has many programs for all ages, including special programs for teachers and students. One week for adults is $350, which includes dormitory lodging and meals.

Whale watchers fix location on whales off the Hawaiian coast. Photo:
Stephanie Ocko

Chapter Five

The Scientist's Experience: Would a Doctor Take Volunteers into the Operating Room?

" "The biggest misconception is that the scientist is going to make a big discovery, that there will be that big 'aha!' on an expedition," says atmospheric chemist Dan Jaffe. "In fact, that is not likely to happen. This is a long-term project."

Jaffe is getting ready to take his first volunteer science team into the Arctic, where he will conduct research on air pollution. For the past fifteen years, scientists have accepted the belief that air pollutants trapped in snow that falls in the Arctic originate in Europe and the Soviet Union. But Jaffe thinks some of the pollutants might be generated closer to home, at Prudhoe Bay, site of a quarter of all the oil production in the United States. To prove this, he needs to collect snow samples that he will analyze later in his lab. To get help and money, he took his proposal for field research to Earthwatch, who submitted it to peer review, approved it, and found him four volunteers willing to go to Alaska in early April and cross-country ski across remote stretches of tundra collecting snow samples.

Working on big projects, scientists focus on ever-smaller components of the work, which often entails going into the field collecting samples or measuring things, interviewing people, or observing animals or insects. Usually, the more data collected by many people, the better. "Ten people in two weeks can make a big dent in a lot of projects," said Ann Post, Smithsonian Research Expeditions spokeswoman.

On a good project, volunteers sense that they are exploring *with* the scientist as he gropes for answers to the problems he is trying to solve. When volunteers arrive on the scene ready to help, very often all they see is the small component. If the scientist fails to paint the whole picture of his research, along with its environmental implications, the volunteer can get lost. "You can get very involved catching little frogs," said one volunteer, "without knowing why you're catching the damn little things. Bridging that gap between the frogs and the overall objective is critical."

Every bit of fieldwork is important; not all of it is obvious. Blue Magruder of Earthwatch says they try to instill that fact in project leaders. "Otherwise, a volunteer comes back and tells his friends he spent two weeks inserting rectal thermometers in musk ox without knowing why."

Jaffe is ready to explain the big picture, but he worries about training his volunteers correctly. He has to make them aware that their responsibility is important because his science is serious, even if what they will be doing is simplistic.

"Even if just one of them never gets it, and collects the samples all wrong, that will be a real drag," he says. "Because then all of the samples will have to be thrown out." Volunteers will do some chemical measurements in the lab. Jaffe will do the heavy-duty chemical analysis later.

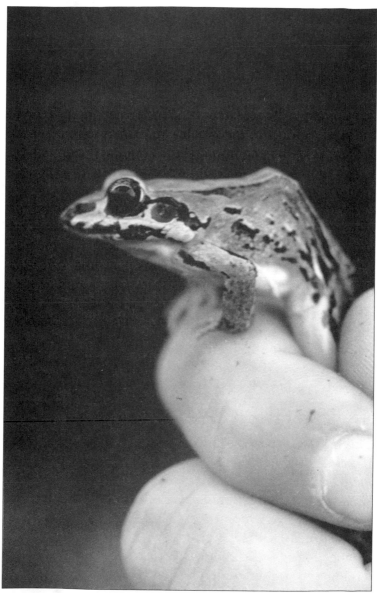

South American tree frog. Photo: UREP

Most project leaders plan a grab bag of things to do so everyone will be busy and happy and feel part of the project. Smithsonian volcanologist Bill Melson uses spry volunteers to scramble up the mountain in Costa Rica and collect lava samples, complementing those who prefer to sit in the lookout station and record the volcanic bursts. Another scientist feels out his volunteers at dinner the first night, mentally placing them where he thinks they will do best. Training in any aspect of a project usually takes about two days. Some scientists complain that by the time volunteers learn what's going on, it's time to go home. Others, like wolf specialist Bill Robinson, like the change. "Every two weeks you get a new crew who are fired up and want to work." Most scientists dream of the three- or four-week volunteer, but not many people have that kind of time to donate.

Getting volunteer funding is like getting any other funding, except it can come with as many as ten people. Middleman organizations screen applications for the validity of the science and the feasibility of attracting volunteers and depend on peer review for final judgments (even if one scientist asked pointedly, "Why are the peers always right?"). After the project, the scientist is expected to publish his results. This may take a while. The volunteer who waits to see what it was he did may wait a few years. At the Foundation for Field Research, the scientist must publish his results in a short form, usually a summary for the newspaper/catalog, then has up to 24 months to publish a longer version, at which time he will receive his final stipend.

One of the responsibilities of the scientist is working out a budget for equipment, transportation in the field, and

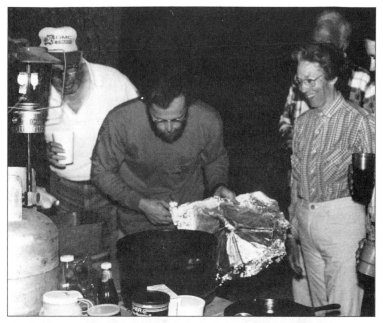

The field manager of an expedition revealing his culinary art: dinnertime on a rock art expedition in Colorado. Photo: Foundation for Field Research

so on, as well as for housing and food for the volunteers and staff. This cost is fed back into the price charged the volunteer, and the honor system prevails. It seems churlish to ask someone who's saving the environment what he pays for rent on your grass hut, but you have a right to know. Ditto, food.

No charm school exists for learning the peculiar knack of being a good tour manager/Eagle Scout/rocket scientist rolled up into one warm personality. Most MMOs require that their project leaders bring staff assistants to handle details and act as a buffer zone. Staff assistants range from dedicated graduate students to spouses to passing paramours. It was for this reason that FFR developed the concept of the field manager, someone sent out from the middleman organization, trained to mediate between scientist and volunteer.

The peer review process usually screens out detail-brooders and great dictators. But funny things can happen on the way to the expedition. Two volunteers remembered an archaeological dig they were happy just to have survived. "The archaeologist was a slave driver. She worked us from 8 a.m. to 6 p.m. I think she just didn't realize that we paid to come here to work, but we don't expect to work ourselves to death. On top of that it rained every night. The tent we were in had one of those little zippered windows that, if you unzipped it, let the rain pour in, and if you zipped it up, turned the tent into an oven."

Home, sweet home. Base camp on a botanical expedition in Anza Borrego, California. Photo: Foundation for Field Research

Another scientist with the personality of a church deacon on this side of the Atlantic became a Lothario on the other side of the Atlantic. In addition to his romantic escapades, he decided to trim his budget by cutting back on food. Hungry volunteers met in secret midnight meetings and ultimately mutinied. Most MMOs do not renew applications from scientists about whom the volunteers complain.

The Field

Anything can happen in the field. Ask any scientist about fieldwork, and he or she is likely to say something like, "No day is like another" or "You have to expect the unexpected" or "Flexibility and adaptability are required in the field." A scientist's nightmare occurred in the Galápagos when a finch expert met his team of volunteers at the plane, took them by truck and boat to the remote island where they would be studying finches, and then found that there were no finches that season because of a dry spell. Not missing a beat, the salty scientist used his volunteers to collect samples of what finches eat.

The weather can bring everything to a halt. Rough seas can interrupt scuba dives and make people on boats wish they were dead. Intense heat affects the number of volunteers able to perform simple tasks; unrelenting rain can bring scientists to their knees; volunteers can get lost in snowstorms; vehicles break down; tools get buried or fall overboard. When computers get involved, the stakes are higher. "We are wholly governed by our technology," lamented one marine archaeologist who was using underwater scanning devices. "I spend most of my days pounding on the equipment and most of my nights worrying." Scientists who do fieldwork develop a belief in Murphy's Law, then worry when it doesn't prevail.

Jaffe worries about the Arctic weather. Last year in March, the nights were minus forty degrees and the days were zero to twenty. This year might be warmer, he thinks, but the steady wind can make it feel fifty degrees colder. Plus, they will spend two nights in a remote field station "150 miles from the nearest anything," he says. "I have to plan ahead. I have a reasonable vehicle, a first-aid kit, and I'm reasonably well prepared. But it's a very serious, remote part of the world." Jaffe speculates that in a worst-case scenario, if someone crawled out to the road and was able to stop a passing trucker with a CB, the area is so unpopulated that response to his SOS would probably take 8 to 12 hours. "I told my volunteers this and made sure they're all in good health."

Juggling research, volunteers, and the field, scientists, like physicians in triage, list priorities. For Jaffe, volunteers have first priority. "I want them to have a good experience. Not fun and games but involvement in the project. Second, I hope we can do some reasonable science." But for long-time whale researcher Adam Frankel, his main priority is taking care of the whales. Frankel is the leader on an Earthwatch humpback whale project off the Big Island of Hawaii. Along with the permit to study whales from the National Marine Mammal Laboratory go restrictions limiting the distance between the research boat and the whales to 100 yards. Whales might violate that distance, but when volunteers do, it's another matter. "One guy came with thousands of dollars worth of underwater camera equipment, expecting to get in the water and play with the whales," said Frankel. A simple mistake by the volunteer created panic in the scientist. On Frankel's priority list, after the whales come the research needs, then the volunteers.

At the end of the project, the project leader will have more data, no matter what. Veteran scientists expect the best and know that volunteers will tend to rise to any occasion, no matter how bad. "You shouldn't expect a perfect expedition where everybody will love it and be interested in your work," said one. "Not all will. But they've paid their money and they should get something out of it." If volunteers want to know why they are collecting frogs, and no one in charge tells them, they should ask.

Without a doubt, at the end of the trip the volunteer will have learned *something* about the environment, about another area of the world, about science, and maybe about himself. As one scientist observed, "What volunteers are like when they come into the project is not that they're like when they leave."

Gear and More Gear

Another aspect of the field project is gear. It has to get there and it has to get back. "I figure I spend a third of the time hauling, fixing, and cleaning the equipment," Frankel said. All projects involve gear, but boat projects seem to involve even more. Between the gear and the weather (it has rained for four days, the worst season in twelve years), not to mention the whales, it is no wonder that everyone says that one day is never like the next in fieldwork.

Could this be an assembly point? Photo: UREP

8:00 a.m. This morning they learn that one of the buoys with a radio transmitter is missing. Did someone take it? Deliberately cut the mooring? How paranoid can they afford to be? "Maybe we screwed up on it somehow," says Frankel. "Maybe a boat accidentally got tangled in the line. Who knows?" From the shore station, Frankel watches as the Zodiac speeds out to where the mooring was. Mike and Tom will dive to try to find it.

The shore station is a simple operation: with a theodolite and two small computers, the volunteer team will fix the whereabouts of the whales, information that is then transmitted to the researchers on the boats. But the road to the hill is steep and strewn with small, sharp volcanic rocks. Six team members make three or four trips carrying the gear: the theodolite in two parts, one very heavy; four metal folding chairs and three folding beach chairs for the computers and computer readers; two computers and batteries; water jugs; and big boxes for stuff—binoculars, personal gear, lunch bags, cameras, a pup tent that will house everything when the sun is unbearable and where the team can eat and cool off. At the end of the morning, the radio crackles and a voice says they've located the buoy, but the transmitter is dead so they're bringing it back for repair. Adam sighs; it will probably take the better part of the night to fix.

5:30 p.m. At the dock, the Zodiac is already in, being angled to connect to the van. Christine pilots the Boston Whaler alongside the pier, while Neuille grabs onto the dock and Mike nimbly wraps a line around a cleat. Now it's time to unload the gear.

Some of it is backbreaking heavy, and some of it has to be collected from the bottom of the boat. From the Whaler, Christine and Neuille heave and haul it up to Mike: gasoline jerricans, metal boxes containing cameras, the video, recorders, boat gear, life vests, fuel lines, the yellow flag that indicates it is a registered research vessel, scuba tanks, masks, flippers, towels, suntan lotion, sunglasses, flip-flops, net bags, lunch bags, the transponder, and the ship-to-shore radio. Some of it has to be washed under the shower at the dock; all of it has to be carried in several trips to the van.

Tom is hosing down the Zodiac; everyone wades through mud puddles to pack the van. The Zodiac crew share their experiences with the Whaler crew, over the noise and merriment of a group of Hawaiian fishermen, who sing and sway under a tree as one plucks a ukelele and another passes around a bottle modestly wrapped in palm fronds.

Then, tired, wet, stiff with saltwater, with muddy feet and dirty hands, the team piles into the van. The van feels clean and warm on the ride home, with the late afternoon sun streaming across the green hills and Eric Clapton playing softly on the radio.

At the house, the whole process is repeated as the gear is loaded into the basement, separated according to boats and shore station, in need of repair or cleaning.

"Everyone's going to the beach," says Mike.

"I have to fix the transmitter," says Adam.

"You've got time for the beach."

"You're right," says Adam.

Chapter Six

The Group Experience: Remember the Donner Pass

"I always wanted to go on one of these things, but I never had the guts. Then I heard about this seventy-year-old woman who went on one in Ethiopia. And I figured, if she can do it, so can I."

—First-time volunteer

"I go because I believe it may be doing some good for humanity. I say that without being naive about it."

—Veteran volunteer

"I've found that most people who go on these vacations are in a state of personal transition."

—Veteran volunteer

"They're already committed because they've paid all that money and gone over there; the question is, how can you utilize them to the extent they're capable of being utilized?"

—Scientist

Airline pilots, architects, stewardesses, plumbers, retirees, postmen, accountants, bank presidents, high school and college students, housewives, gamblers, photographers, carpenters, small business operators, artists, cartoonists, lawyers, ministers, psychiatrists, musicians, economists, surgeons, librarians, teachers, truck drivers, stockbrokers, writers, chemists, engineers, computer specialists, ex-CIA operatives, physicians, layabouts, and adventurers are some of the people who are environmental vacation volunteers. What this means is that anybody can go on an environmental expedition who has the motivation and who wants to spend the time and the money. The only restriction is age: you have to be at least 16 on most projects; 18 on Habitat for Humanity International projects; 13 to 14, or "mature," on Foundation for Field Research projects.

"Right away you size each other up," said one volunteer, "and there's always one or two who want to run the whole show, so you might as well get used to it right away. Of course, I'm always the perfect one. Just kidding."

Whatever motivates anyone to go to X and do Y contributes to the overall chemistry of the group. Fate brought you together: keep that in mind if you look around the first day and wish you were back home. Blue Magruder at Earthwatch says they have no way of knowing what a volunteer's secret agenda might be. A less psychological rule of thumb is that in any group of fifteen, three are likely to share the same astrological sign. The discovery process of ten total strangers on a labor gang can be a lot of fun. Many a wild animal must have paused in the bush to wonder about the distant chatter extolling the virtues of California or criticizing the Mets.

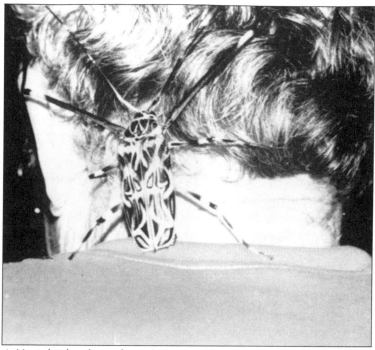

A Hercules beetle explores its own territory. The brave volunteer is on Trinidad. Photo: Foundation for Field Research

Most groups bond as a whole because they must stick together to face an unknown environment and a new learning experience that involves actually using what is learned. "It's not easy, but anyone with a little intelligence can do it," a volunteer observed. Project leaders are prepared by previous scientists and MMO staff members to be aware of the rise and fall of interest levels. The initial burst of enthusiasm on the learning curve drops off at the end of the first week when the high romance dissolves into routine. It stays there until the end of the second week when enthusiasm rises again, so that by the time the volunteers are ready to go home, they are experts in the unusual, from measuring giant clams in three meters of water to spotting fish bones in wolf spoor.

By that time, they have entered into an odd brotherhood. Back home, they find themselves strangely alone. "The problem is, this type of experience is so far removed from the common experience of your friends that they can't relate to it," said a volunteer. "Here I am showing these fabulous slides I've taken from the other side of the world, and I look around the room and everyone's falling asleep."

A first-time project leader who discovered that several of his volunteers had previously been on a number of other projects described them as "this secret subculture" of people with unusual skills. What is developing around the country, said a spokeswoman at UREP, is an increasingly skilled cadre of volunteers. "That's why the returnee is so important. We have some people who only go on archaeological projects, and they bring real skills," she said. At the Smithsonian, project leader Bill Melson is proud of his "veteran volcanologists" from his Costa Rican volcano project.

From the scientist's point of view, the problem volunteer is the one who doesn't want to work. "I've had some who wanted to be wet nursed," said a veteran project leader, "and some who were just looking for names to drop at cocktail parties."

The only mistake a volunteer can make, according to several scientists, is not being culturally aware before going into the field. It is one thing not to read the science literature, but it is far worse not to know something about the country you will be working in. One scientist remembered

Volunteers photographing a large sea turtle laying eggs on the beach in Michoacán, Mexico. Photo: Walter Meyer

a volunteer on a project in French Polynesia who asked, "Why is everything in French?" Because he found the place "unsanitary," the scientist said, "he drank sodas and ate cookies and left after ten days."

Margaret Harvey, coordinator for Habitat for Humanity International, said knowing something about the culture of the host country is essential to any kind of helping project. It is not always necessary to speak the language, because translators are provided in sensitive situations (the only exception is Amigos de las Americas, below). But being aware of the cultural differences is the first step to understanding the value of the similarities. IRE's director, Mark Richman, said, "Other cultures can be alien environments for some Americans." Extensive poverty, obvious disease on city streets, and begging children in rags can be unsettling.

There is not much information on the psychology of short-term adult work groups in far-flung places, but middleman organizers recognize that the worst thing a volunteer can do before he or she goes is to fantasize about what it will be like. "Bring as few expectations or preconceptions as possible," advised a scientist. "Don't imagine it's going to be like the color brochure," said another.

The gap between fantasy and reality can be paralyzing. One volunteer, having traveled literally halfway around the world, retired to her tent the first day and refused to come out until transportation was arranged to take her back to the airport. Another volunteer, on a project in West Africa, froze with culture shock when she encountered the latrine. She left early.

Sometimes the chemistry is off, or the stars are out of

The buckets contain baby sea turtles. Volunteers working during the Michoacán research trip sponsored by the Foundation for Field Research. Photo:
Walter Meyer

whack. On an expedition to Tahiti, two volunteers decided they didn't like each other on Day One and came out verbally slugging each morning. Finally, the rest of the team demanded they call a truce so everyone could get on with their lives. One scientist remembered a volunteer who was undergoing a "healing process" for an unexplained emotional problem. She took off for three days to be alone, leaving the project leader frantic. She ultimately returned, but the scientist said it was highly disruptive. "We never knew if she'd be there in the morning." But, by and large, tensions are short-lived.

The Assembly Point

As scientists have the field to worry about, volunteers have the assembly point. If anything can go wrong, it's usually at the assembly point, or staging area, where the team members come together with the project scientist in the field. Few assembly points can be found on maps; they *feel* very abstract and read like stage directions in a play: "The project leader will pick up volunteers at the Kathmandu airport between the hours of 1 and 2 on Tuesdays and Saturdays. The project leader will carry a sign."

Typically, this is all the information a volunteer will receive. The directions sound simple enough, but for the volunteer who travels halfway around the world over mountains and oceans making three or four connections, arriving at Kathmandu airport between 1:00 and 2:00 on *any* day is akin to a hole-in-one.

When it does work, of course, it's a thrill. The volunteer's plane touches down with just enough time for him to fetch his bags and arrive in the waiting room, where the project leader bellows a hearty welcome, packs him safely into the pickup, and speeds off into the outback.

But it doesn't always work that way.

The worst case happened in Central America. The project leader, arriving a day before his expedition was to begin, was mugged and robbed of the project's entire budget. Not wanting to test his faith in mankind, he caught the first plane back to the United States, leaving his two volunteers who were en route to the assembly point to fend for themselves. Being resourceful, they finally took a taxi to the place they were to be staying only to find themselves alone. A series of calls to the frantic MMO revealed that the project was canceled. Their money was refunded, and they were now tourists. Fortunately, they were in a place with archaeological and wildlife wonders.

Another volunteer missed his connection in Chicago, got backed up in Los Angeles, and arrived a day late and two minutes short of midnight in a little airport in Peru. The only other people in the airport were begging children. When he gave them some coins so he could try to catch some sleep, they ran off and came back with their whole family. At dawn, the volunteer phoned the project. Then, hanging on in the back of the hay wagon, which the family insisted he use as a taxi, the tardy volunteer shouted out the directions and guided his driver to the project site.

On a project in the South Pacific, an MMO bungled its information and failed to inform a volunteer that the assembly point had been changed. Through flight arrangements made by his travel agent, he spent the first night in the airport in the wrong country, before taking a plane to what he thought was his destination. Faced with another

Marketplace, South American style. Photo: Molly O'Malley/IJRFP

homeless night, he appealed to the Minister of Lands who, unable to find the project, invited him to spend the night at his home. The following day, through numerous telephone calls, not easy in developing countries even at ministerial level, the volunteer finally located the island where his project was and got the plane to take him there.

Things happen.

"Anywhere in the world, if you have a problem, someone is going to help," said Robert Schilling, a veteran of "thirty, maybe" science and adventure expeditions. "When you think about it, people are people all over the world. Many want to be good to you."

If you are physically tired, a project like this might not be a good idea, because it will be exhausting. But if you are suffering from psychological burnout, those who know say there is nothing better to restart the fire. Helping someone else do fieldwork provides a sort of mini-workplace experience. It has the structure of work, but it's not the work that your life depends on. Because of that, it's free of complexity and frustration. And, it's a relief from your normal life. As one volunteer mused, "Whatever this is, it is not reality."

Based in a more manageable, less complex work context, environmental vacations do provide challenge: socially orienting to a group of strangers, learning new skills, adjusting to a different sleeping and eating routine, often in a challenging locale. Mastering these challenges provides an enormous sense of satisfaction, or what a volunteer called, "the rush, the high." Added to this is the sense of purpose of doing something on a small scale as well as doing "good" on a large scale.

In a cave in Oro Grande, New Mexico, two AFAR volunteers get into the dirt on an archaeological dig. Photo: AFAR

Chapter Seven

Stripped for Action: Which Way Do I Go?

How can you know which environmental vacation would be the most enjoyable and rewarding for you? One way is to talk to someone who has done it, but even then, what is right for them might not be right for you. Put yourself in the picture: do you enjoy spending hours underwater trying to make sense of a reef that keeps changing every time you move? Is sitting inert for hours in a jungle to chart the movement of one insect something you would enjoy? Do you mind living in a primitive village in a hot and steamy climate where work begins at dawn?

What follows is a selection of typical environmental volunteer projects, with a description of what they involve, what tools you will be using, and how others have experienced them.

Archaeology
Archaeologists are detectives. Tracking clues, they deal in human activities and human interaction with the environment. If a stone knife changed shape, for example, maybe it was because the game that was hunted changed, say, from mammoth to wild goats. Archaeologists start with a theory

—they say that you can find what you're looking for if you know what to look for—but they are flexible enough to alter the theory if they find something that convinces them another possibility is more likely. Like most scientists, archaeologists are always asking why.

There are opportunities for working on digs almost anywhere in the world. Prehistoric archaeology means working with artifacts that come from a culture without written history. That can mean 100,000 years ago or as few as a 1,000 years ago. Historic archaeology dates from the earliest deciphered writing to garbology, which is the archaeology of modern trash.

Marine archaeology projects, in spite of their glamour, do not often use volunteers, because underwater archaeology is a demanding discipline that involves the precision of land archaeology as well as the pressures of working underwater. Marine archaeologists usually use volunteers for survey, photography, and retrieval work underwater.

Most digs reveal clues about past problems that might have value in solving present-day problems, for example, drought in the Southwest. Many of the solutions ancient people found, without benefit of high technology or global communication, have value today.

Archaeology is labor-intensive. Expect dirt. Archaeologists think geometrically: they will teach you to lay out horizontal grid squares that divide the area to be dug into workable units that also correspond to graph paper, on which "finds" can be charted. As the dig descends into the earth, the vertical wall layers, or strata, will show, like shelves in a bookcase, climate and habitation changes.

Few archaeological digs hit pay dirt on the first strike,

and most do not produce gold. Indiana Jones notwith-standing, few archaeologists actually go after the Holy Grail. But that spirit exists, and most volunteers feel the tingle of beginner's luck that might mean something fantastic will turn up.

Archaeologists are also precision oriented, for one good reason: digging destroys the site, so they have a single chance to get it right. "Archaeology is a functional autocracy," said one archaeologist. Archaeologists are used to giving orders to graduate students and insist that all finds must be properly recorded. You cannot just pull something from the ground. Diggers collect bucket loads of dirt from the layers with artifacts. The dirt is then sifted for tiny things: pollen, insect carcasses, shells, anything that will give clues to the environment, climate changes, and what the people ate and did.

Diggers use shovels, trowels, and brushes, and if the digger reaches a skeleton, he will use increasingly smaller brushes and, ultimately, a toothbrush. Expect to haul dirt in buckets and to spend a lot of time on your knees. You will probably make labels for artifacts; you might wash the finds in buckets of water and catalog them according to type; or you might work the sifter. Some days you will uncover nothing; other days artifacts will be found everywhere. But if you are the first person to uncover something dropped 2,000 years ago, it can send a shiver along even the most jaded spine.

Survey archaeology is basically walking and looking. Often archaeologists guess where sites are, usually based on logic: the area has a lot of drinking water nearby, or it is an easily defended site, or there is a mound. Volunteers form

A group of volunteer archaeologists digging on Easter Island. Photo: Sue Ann Wolff

a line and walk a certain distance, noting anything they see in front of them.

In Hawaii, for example a particularly difficult survey covered a forested hill area (which later turned up abundant evidence of early habitation) as part of archaeologist Paul Cleghorn's project. Earthwatch volunteer Phyllis Hoo describes it:

If a person doesn't know how to use a compass, it can be traumatic. The project leader says, "Okay, go straight ahead, south, for a hundred yards into the forest to find anything." You're walking, and looking, going straight, climbing over rocks, going around trees. Pretty soon you wonder, am I going straight? you look at your compass, and you see you aren't going straight. You turn around and you get lost. The forest is so dense everything looks the same. The compass is the only thing that gets you back.

Hoo also became proficient in the use of a machete, which she carried in her belt when she wasn't slicing and dicing her way across the forest. "I knew absolutely nothing about archaeology before," she said. "It's amazing what you learn. Clues are everywhere if you just know what to look for."

In Europe and North Africa, volunteers can join archaeological expeditions that dig Greek and Roman sites. This involves excavating not only items from daily life and skeletons but also sculpture and buildings. Classical archaeology —Greek and Roman—is among the richest in terms of finds and clues about what life on all levels was like 2,000 to 2,500 years ago. Painted pottery, extensive mosaic floors, metal jewelry, elaborate tools and weapons are trappings of the rich. The less fortunate had a different cultural expression, and an ongoing Earthwatch dig in ancient Carthage in Tunisia, for example, is piecing together life-styles of the working class.

The Classical world also had its own cultural outback, areas on the fringes of the ancient centers, along the boundaries of the trade routes. Sardinia, a big island off the Italian coast, above Sicily, is the site of some 7,000 mysterious stone-built towers, dated from 1800 B.C. to about 250 B.C., whose meaning has eluded archaeologists for centuries.

Here, Leonore Gallin of the University of California, Los Angeles, has taken Earthwatch and UREP volunteers each summer for the past ten years to help in her explorations of the mystery of the *nuraghi*, as the stone towers are called. Gallin's teams have made extensive measurements of many nuraghi in an effort to collect data that might indicate who built them and why. They have also conducted archaeological digs around and near the nuraghi.

Not all nuraghi are intact. Some are piles of fallen stones; others are half-ruined; some are tall and imposing. Most have two stories, each with a dome-ceilinged inner room, a stairway built in the wall, and one entranceway. Inside, they are dark and strange, lit only by light coming in through the single opening; and they smell of animal droppings. Many farmers have traditionally used nuraghi as cow, donkey, and sheep shelters, and volunteers working on the inside often find they are knee-deep in tradition.

In the early morning light, across the empty fields, nuraghi glow orange with the lichen that covers them. Teams set off shortly after dawn and carry surveying equipment—meter sticks, theodolite, transit, tripods—as well as tools like scythes for clearing away the thick growth that often blocks the entrance. They trudge over walls and through fields, past the ubiquitous flocks of sheep and the lone donkeys. While some members clear away the tough grass and bushes at the base of the nuraghi, others climb up to the roof and drop a tape measure with a plumb bob attached over the edge to try to gauge the height, often difficult because the roofs have deteriorated into uneven blocks of stone. At the base, other team members work in pairs, as one holds a meter stick and another sites it with a transit or theodolite, to measure distances and to calculate angles. Teams work the same way inside, and, standing on a wobbly ladder supported by two or three people, an assistant or a volunteer will hold up a long meter stick and try to measure the height of the center of the dome. The domes are amazing structures, built with unmortared blocks of stone laid in ever-decreasing circles to the apex.

Did Bronze Age or Classical people *live* in nuraghi? Were they workshops? Why are they all built alike, with individual differences? What happened to the people whose culture they belonged to? Gallin welcomes volunteers and their ideas on her project.

Ornithology

Although bird projects differ, depending on what the researcher is looking for and the habits of the birds, they usually involve temporarily trapping the bird or birds to band them or to study them; sometimes volunteers are asked to count the number of birds in a flock.

FINCHES

In part of his lifetime of work studying Darwin's finches in the Galápagos, scientist Robert Bowman has used Earthwatch volunteers in a number of ways to collect data. Although their variation is greater than many other species, finches are tiny. Trapping them involved stretching out a very fine net between two poles like a volleyball net. Unable to see it, the birds would fly into it and become trapped but not harmed. Volunteers then could measure the birds before setting them free. Using an electronic instrument designed especially for his research, Bowman would hold a finch until it bit the tip of the instrument. A small com-

Handling birds like this marine bird in Alaska requires a tender touch, which this volunteer obviously has. Photo: D. Costa/UREP

puter recorded the kilogram forces of the pressure of the bird's bite. From these data, Bowman was able to extrapolate what kind of seeds the finch was adapted to.

Another objective of Bowman's research was to understand bird song. Volunteers used shotgun and parabolic microphones to record individual and group songs. But in the wild, where finches fly around like presents out of a piñata, the most effective way to get the birds to sing is to play a tape of finches singing. The birds that hear it stop whatever they are doing, start singing, and fly directly onto the mike.

Like most other species in the Galápagos, the finches there are unusual. They have not yet learned to be afraid of humans. As a result, they hop on people's shoulders, stare into cameras, and are as curious about people as people are about them.

RAPTORS

Raptors, or birds of prey, are as large and languid as finches are tiny and busy. In the Carrizo Plain, near Bakersfield, California, red-tailed hawks cruise in the sky with barely a ripple in their wings, their sharp eyes watching the ground for any sign of movement of the small animals they swoop down to catch and eat. Unfortunately, too many hunters have taken their toll, and these birds are now endangered. By briefly trapping them, researchers, with the help of volunteers, can band them to better understand their numbers and territories as well as their health and growth rates.

On a weekend crane and raptor count with FFR researchers Bill Everett and Sherry Teresa, volunteers slowly walked and drove across the flat stretches of the Carrizo Plain looking through binoculars and spotting scopes for the few red-tailed hawks that live there. Volunteer Claudia Cline describes the experience:

> The birds of prey are fantastic. Just the hunt for them, driving around and looking for them, was exciting. Then we put out a small wire trap with two live mice inside. On top was fishing line with open slip knots. When the bird sees the mice, it dives and lands, and its talons get entangled in the knots.
>
> Then the researcher takes hold of the bird, gently takes the feet out of the knots, and puts the bird on a towel on the ground. Then everyone goes to work, measuring the wingspan and links and putting a band on it, before setting it free.
>
> We were actually able to hold a wild red-tailed hawk in our hands. We held it around the feet and neck, keeping our hands away from the talons. It was just a thrill to be that close to a wild bird and to be able to see its beautiful plumage.

Paleontology

If our country had a national animal, it would probably be the dinosaur. Dinosaurs inhabit science museums and playgrounds, star in cartoons, cling magnetically to refrigerator doors, and inspire scientists to hold heated arguments about the cause of their apparent quick and complete demise. Teachers find dinosaurs exciting because they are a good model for evolution: dinosaurs arose several hundred million years ago, diversified into hundreds of species, then vanished. Whatever caused their end, dinosaurs either did not have the time or the ability to adapt. Could the same thing happen to us?

On a plateau in Arizona at an elevation of about 6,000 feet, a few miles from the Petrified Forest, hundreds of specimens of creatures that predate the big dinosaurs lie buried. This is where Kevin Padian, a world expert on flying reptiles, has chosen to do fieldwork with the help of UREP volunteers. The trek into the outback is a trek into the Late Triassic, about 220 million years ago, Padian says.

It's a very important time in the history of vertebrates, because there are three kinds of animals: a bunch of old things just having their last gasp; then there's a radiation of really weird things that only show up in the Triassic; and then at the end of the Triassic, there's the beginning of what turns out to be the whole modern fauna today: mammals, crocodiles, turtles, amphibians, snakes, plus the dinosaurs and terrasaurs. It sets the stage for what's going to happen in the next 250 million years to the end of the Cretaceous, and for 6 million years after that.

Given these possibilities, Padian's volunteers fan out across the desert and "prospect" for old bones. But one day it rained, which forced them away from the original, established site, and they trudged through desert mud for a mile in the other direction to a new area. "It looked like a rock. I squatted down, then I recognized bone, and bone up against another bone, until I realized it was an entire skeleton still together," says Steve Bailey, a volunteer. "I kept brushing away the earth, which was like eroded concrete, from the bone, and I uncovered more and more. It was unlike anything I have ever experienced."

What Bailey had uncovered was a complete Phytosaur, about 20 feet long, weighing about 200 pounds, and about 220 million years old. For a week, half the team of sixteen worked on the delicate job of extricating it, by brushing away the surrounding dirt, then packing it with newspapers and tissues, before encasing it in plaster-soaked burlap in four parts. The skull itself was a meter long. Then came the problem of getting it out. The nearest road was more than a rugged mile away. One of the neighbors, a veterinarian, brought in his 4-WD pickup, and team members hauled one of the parts into it. But the next day, after waiting for him to come back, they learned he had to be in a parade in nearby St. John's. "That's when Jay, one of the volunteers, took it upon himself to drive his brand new Ford truck across the desert," said Bailey. "We were screaming at him to stop, and he says, 'No, I'm going to do this! I owe it to the group!' And he did it."

Bailey, who is a science teacher, described the Phytosaur as looking like "a low-slung crocodile-shaped creature with big skeletal differences from crocodiles." Padian is more direct: "It looked like a big crocodile with goiter." Whatever it resembled, it shortly became extinct. The earliest dinosaurs, Padian says, were "fairly small and not very numerous. They don't really get going until the other things get out of their way."

Volunteers also found Phytosaur teeth, which are "serrated like steak knives," Bailey says, and parts of other animals. They picked them out of the sand, cleaned them off, prepared them, and in some instances, got them out of

Volunteers weighing and measuring a loggerhead turtle at a research station in Baja California, Mexico. Photo: Walter G. Meyer

Volunteers participating in the excavation of a Roman burial ground in Germany which dates to A.D. 100. Photo: Foundation for Field Research

Researcher studies langurs on the slopes of Mount Abu, Gujerat, India. Photo: Dan Hrdy, Anthro-Photo

Researcher recording elephant seal sounds on the California coast. Photo: Burney LeBoeuf, Anthro-Photo

Members of an expedition recovering Bronze Age artifacts in Lac de Neu-
chatel, Switzerland. Photo: Dwight Sieggreen

Diving for sea sponges off the coast of Fiji. Sponges were being gathered to evaluate their potential medicinal qualities. Photo: K. Granger

Habitat for Humanity house-building project in Boston. Photo: Michael Price

Dr. Daniel Anderson and a UREP team member band endangered pelicans in Baja California, Mexico. Photo: Jean Colvin

A murre chick relaxing after being banded on St. George Island, Alaska. Photo: Dwight Sieggreen

A team of volunteers working with an archaeologist documenting rock art in the American Southwest. Photo: Foundation for Field Research

Volunteers participating in the excavation of a medieval castle in Wales. Photo: Foundation for Field Research

UREP Director Jean Colvin with two Rendille men of northern Kenya. Photo:
UREP photo file

the rock with plaster jackets. "That's real messy, and lots of fun," says Padian. "It involves getting real dirty and wet and full of plaster and toilet paper."

The days are hot. According to Bailey,

> Our little pleasures included someone saying "Oh, the wind is blowing." The typical prospector-fossil-finder's lunch included a can of fruit cocktail and a can of stewed tomatoes. It sounds disgusting, but it was glorious out there, because it was wet.
>
> One day one of the local teachers brought her students out on condition that they each carry one or two gallons of water. We needed the water, to drink and to fix our plaster with. Imagine these little schoolkids trudging about a mile across this hot desert sand. They were just completely spent. But when they came over the rise and saw the pit, they were refreshed again, and they started asking great questions.

Padian says the townspeople are delighted they are there. "They love learning about their fossil heritage. They bring their kids, relatives, video cameras; go all out."

Accommodations in St. John's are modest. "We stayed at a little cheap motel at the group rate," Bailey says. "The motel served us a standard breakfast at 6:00 or 6:30, then we'd walk that mile, carrying all the gear we'd need. It was really hot. Then at the end of the day, we'd pack up, go back, get cleaned up, meet for dinner, talk about things and enjoy the evening. From the plateau, we'd watch the thunderstorms form. It's as dramatic as can be, with dust floating around, lightning flashing. Because it's so high up,

Volunteers making observations during a UREP-sponsored botany expedition. Photo: UREP

we could watch clouds forming as the humid air coming up from Mexico met the plateau. Just the lightning flashing out of the clouds was dramatic; then it would rain a few huge raindrops.

"When I was a kid, I always hunted little shell fossils and arrowpoints, and I was a kid again here. When we got back, my students asked, 'How much did they pay you to do this?' And I said, 'No, no, no, *I* paid.' And they said, 'What an idiot!' And I said, 'Well, you don't understand, kids.'"

Steve and Leslie Bailey can't go back this season because they are expecting a baby. Another couple from last summer's paleontology project is getting married. Padian claims that both those events are a result of his project. There must be more to dinosaurs than meets the eye.

Other paleontology projects in Mexico and Montana look at very early ocean creatures and the later huge mammals that migrated from South America; another tries to understand what happened during the Cretaceous when dinosaurs disappeared. In an old hot springs pond in South Dakota, Earthwatch volunteers help dig out parts of more than a hundred mammoths.

Botany

In the tropical cloud forest of the rugged mountains of Ecuador, UREP botanist Grady Webster has begun to inventory all the plant species in the Maquipucuna Tropical Reserve. "The real problem is that Ecuador has at least as many plant species as the United States and Canada put together. And it's a country no bigger than California," Webster says.

With assistants and xeroxed informational sheets that

give an idea of what to look for, Webster leads his volunteers on newly cut trails through the cool dark forest, under an umbrella of trees 30 to 40 meters high. Ferns and orchids are everywhere. Volunteers place the specimens in a plant press and fill out and attach special labels. At the end of the day back at the lodge, they write them up in a record book, dry them ("with two dryers working off butane"), and then Webster takes the specimens to Quito.

"I'm not a botanist," says volunteer Molly O'Malley, "but I learned more with Grady in three weeks than I ever did before. It was rare if he didn't know every plant we came upon in the jungle. If he didn't, it was even more exciting, because it might be a whole new species." Webster so far has collected only a few hundred species but expects to find at least a thousand.

Volunteers stay at a rustic research station lodge, with a roof, a floor, and three sides, and beds with mosquito netting. A nearby river flows into a mountain pool where volunteers can swim. On free days, Webster has organized a trip to a volcano, where hardy souls climb up to the glacier at 18,000 feet. But the lodge, Webster explains, is right at the level of the clouds, which means that as the air rises up from the sea, it condenses, and if it doesn't rain, it is very humid and cool at night.

When O'Malley got back to the third grade class she teaches, she replicated her own experiences in Ecuador. Her students collected plant species in the San Francisco area, pressed them, dried them, and documented them. They compared Ecuadorian plant species, which O'Malley had brought back, with plant species in a local park; and they visited the cloud forest exhibit in the San Francisco

Arboretum. Her class got so excited that they built a rain forest in their room and cried when they had to dismantle it at the end of the school year.

To help him with his enormous task, Webster collaborated with a university in Quito to have an Ecuadorian botanist on his staff. "It's very important, especially in Latin America, to get more people involved. I hope they'll continue plant collecting," he says. "It would be nice to motivate someone there." To try to inspire them, he has invited the local townspeople to join his project for a day and do what the volunteers do.

After the project, O'Malley went to the Galápagos, which is easily accessible from Quito, and traveled in Ecuador. She found what she thinks is the least-known bargain in the world: a hotel that charged 85 cents a night. "And it had water!" she says.

Most botany expeditions collect, press, and record. But on Irwin Ting's UREP project on the island of St. John in the Virgin Islands, volunteers take turns staying up on 6-hour shifts through the night to monitor instruments attached to succulent plants that measure the plants' physiology. Ting's long-term research with desert plants and cacti has led him to study succulents in the lush forests of St. John to try to find out if in addition to reducing their water loss during daylight, they also reduce it at night. "If we can tell when they're losing water, we can tell when they are resistant to drought," says Ting. With the world's water resources threatened by overuse, understanding drought resistance in plants is of primary importance.

But succulents are not the only drought-resistant plants

on St. John. On one small island, Ting can study epiphytes and the strangler fig tree, which seems to exhibit the same kind of drought resistance as succulents do. St. John is a national park. "That means that when we go back every year, the same plants will be there. They won't have been replaced by a parking lot," he says. Ting's volunteers tag the plants they identify to measure annual growth.

Because the island is small," he says, "it's possible to look at all kinds of plants." Walking along trails, volunteers count the clusters of succulents, the small epiphytes that grow on trees, and the huge strangler figs. From that data, maps can be made, and Ting can see how drought-resistant plants correlate with water availability. "It took us six to seven hours to cover one trail three miles long," says Amy Ballin, a volunteer. "It's not physically hard, but you have to walk really slowly. It takes about a day to recognize the plants you're looking for. Then you get very good, and by the second week, you have it down perfectly." The island of St. John is small, with some 1,300-foot hills and 30-degree slopes.

At the end of the day before the night teams will set up their equipment, volunteers meet with Ting and discuss what they found. They are encouraged to give their own points of view. Ting likes meeting with the volunteers, and believes they contribute a great deal, even if they have no scientific background. Ballin says that the best part is the feeling that they're all in the exploration together. "One of the fun things is that, because the research is ongoing, Ting doesn't necessarily have all the answers, and he's interested in what *we* think about the research."

"It's best to be an outdoors person, or at least in reasonably good condition to come on this project," says Ting. But then he remembers a volunteer who had lost the use of his leg muscles after an attack of polio when he was a child. "He walked on crutches and rode in a jeep," Ting says. "He got on fine."

Entomology

Scientists find hundreds of new insect species every year. How many remain to be found they cannot begin to calculate. If the numbers don't humble researchers, their complex diversity does. Insects are adapted to their environments not only in camouflage but in defense mechanisms. Joan Thuebel, a volunteer on Dr. David Nickels's katydid project in the Amazon, recalled seeing what looked like "a fairly large roach. But when the researcher poked it, it sent a puff of smoke straight up from its back to frighten predators. And if that doesn't work," she said, "when the researcher poked it again, it shot out a purply-brown liquid that hit his hand."

Most botany and entomology projects take place in rain forests, and most rain forests are endangered because population pressures have led locals to cut down trees either because they need the space for agriculture or to grow something for export, like palm nuts or oil. Rain forest preservation projects depend on public awareness, and, as Harvard entomologist Gary Alpert says, it's easier to interest the public if elephants or monkeys are endangered. Ants—Alpert's specialty—don't inspire too much empathy, even if their numbers are an excellent indicator of the biological health of an area. "Ants are good bioindicators. Their habitat diversity and richness and numbers of species imply other diversities," he says. Lots of ants means lots of bugs, which means lots of frogs and chameleons, and so on, up the biological chain. Alpert studies ants in Madagascar, an island off the coast of East Africa. Its isolation, says Alpert, has allowed a lot of biodiversity; all kinds of life forms have evolved, including insects found nowhere else on earth. But because the rain forest is being cut down to make way for rice fields and for palm trees to create an industry for palm products, the whole area is endangered. "There's an urgency to see what's there and to protect it." But the protection of the rain forest is most effective if it is administered on local levels. "A scientist can't do it alone," says Alpert. "It has to be done on a grass-roots level."

Still, the mere presence of a scientist in a rain forest can have residual effects on its ultimate preservation. "Most governments don't look at rain forests as having any value, apart from what's being cut down. But if a scientist comes in, suddenly they think maybe the rain forest is valuable. If the scientist meets with officials and trains local scientists about life forms there, that can pique the interest of others," Alpert says.

What volunteers experience on entomology projects in rain forests is the incredible *connectedness* of life. Mrill Ingram, for example, a volunteer on a project studying ant behavior in Costa Rica, wrote in *Earthwatch* magazine (April 1990):

> Most exiting to me at this point is my dawning awareness of the intricate connections between all these creatures that weave them into one great relationship called a rain forest. I've seen many of these animals

before, in zoos or on film, and I am convinced that what makes them so awe-inspiring here isn't just their freedom; it's their business. There are sophisticated connections among the seeds in front of me, the ants crawling nearby, the birds overhead, and the trees towering above me.

When volunteer Thuebel left Equito, Peru, and headed down the Amazon in a long, thin boat with a weak motor, she knew only that she would be studying leaf-mimicking katydids. The first few nights of her Earthwatch expedition she and her team stayed at an inn with running water and electricity. Then they headed deeper into the jungle, up the Snapu River, a tributary of the Amazon, for three hours. Base camp there was a platform on stilts, with a high thatched roof and open sides, with mattresses covered with mosquito nets. There she and her teammates discovered that the jungle at night is louder than the Santa Ana Freeway.

Volunteer Rob Pudim recalled an Amazon night: "Strange sounds that seemed far away but right up close. Weird echoes of things that deceive the ears. One night our team was walking along a trail outside the camp when we all heard a party going on, with people laughing and talking, glasses clinking, and a sort of whacked-out music and drumbeat fading in and out," he wrote in *Earthwatch* magazine in March 1989. "But these are auditory illusions. There are crickets that sound like someone tapping an empty glass with a spoon. Then there are the mating calls of a thousand crickets, grasshoppers, and katydids—a racket of dry clicks, whirrs, buzzings, raspings, ratchetings, high-pitched trills and keenings."

In the morning, Thuebel's team put blotting paper saturated with 12 different orchid essences on 12 different trees, then sat down to watch and wait for bees. The object was to get a sampling of Amazon bees and which orchids they preferred to compare with bees being studied on another project in Central America. When a bee landed on the paper, volunteers would scrutinize it to see if it was scraping it with its front feet and appearing to transfer the essence to its back feet. "That meant it was serious about the scent," says Thuebel. "Then a volunteer would take a butterfly net, sneak up to the tree, and very, very quietly swish the net and grab the bee in the net." Then came the fun part: getting a "bee that's as mad as a hatter" out of the net and into a jar of cyanide. "Don't worry," the researcher told them, after saying the only way to get the bee out was to reach in and get it, "only the females sting. The males don't, and they are the ones who are attracted to the pollen." The procedure worked well until the last day when a volunteer got stung. It might have sent the researcher back to the lab, but for the volunteers, the answer was simple: a hermaphroditic bee, says Thuebel.

The team spent the afternoons gutting and preserving the bees for transport, carving out their insides with a tiny Exact-o knife, then stuffing them with cotton and putting pins through them. Thuebel described it as "picky, picky little work. That's part of the scientific process. Some parts are thrilling, and some parts are very boring."

The thrilling parts came at night. The katydids, between one-half inch and two inches long, mimic their leaf homes so well during the day they are invisible. Between 10:00

p.m. and midnight, the team, carrying large flashlights, walked into the jungle to find them. "At night you can see them move, because they are out to eat and to mate," Thuebel says. Pudim describes the night jungle: "The air is thick and motionless. No stray breezes here, not even a faint movement of air to stir the hairs on your arm. It's like no other forest on earth. An adrenalin surge like you feel in a dark, musty cellar when a cobweb touches your face is triggered by the whispery, tickly brush of wings on your face. Probably a moth. Or a bat. Probably."

The purpose was to discover the varieties and adaptations of katydids, and as many as possible were captured and preserved. The team found 13 new species in two weeks.

Back home, Thuebel found she still swats flies and mosquitoes, but she has a new respect for things she doesn't recognize. "I have a whole new attitude. I'm concerned about the environment," she says, "and this is one way to contribute. I'll always be on the periphery of science, but now I find little things that fall into place. For example, learning that the poison-arrow frog, which I saw, is being used in research for a cure for Alzheimer's disease. Now I'm aware that if we don't go out there now and learn what's there, we may never know."

Anthropology

Just as species become endangered, so do cultures. Culture can be fickle, reacting to changes within groups, and sometimes when a part of a culture is lost, what replaces it is not as valuable. For example, UREP's director Jean Colvin witnessed the disappearance of the traditional, finely woven

Andean highland couple in festive dress, Bolivia. Photo: Cynthia Le Count/UREP

root fiber water baskets of the Rendille people, camel nomads living near the Ethiopian border of Kenya.

"In every household ten years ago, there were one, two, sometimes three. They took a long time to make and were woven so tightly that not a drop of water seeped out. Then came the cash economy, and things changed," says Colvin. A video made by a UREP team recording activities near a well shows women arriving on camels, bearing bright orange, green, and yellow plastic jerricans. The woven water baskets no longer have any value in the culture.

Recording lost cultural expressions—everything from

household items to tools, toys, and musical instruments— requires writing down all the ways in which an object was made and used. Among many tribes around the world, possessions are special, and tools have their own "souls" that translate into power for the user. This is the kind of thing that gets lost when jerricans take over.

"Interviewing takes a lot of patience," Colvin admits. In Tanzania, UREP sponsored a project in a remote village in which every item would be cataloged and samples sent to a museum in Tanzania. "The process involved sitting cross-legged or on your knees in a hot, smoky hut for four hours or more, talking about household items. Afterward, one of the volunteers asked, 'How long can you talk about *spoons?*' Once you get used to it, it's fun. There would be no need for archaeology if we had these kinds of records," says Colvin.

On other MMO environmental vacation projects, volunteers have interviewed Greek women about their traditional costumes, Highland Bolivians about theirs, and itinerant tinkers about their unusual lifestyles. The process usually involves going into homes, establishing a relaxed, social atmosphere, and talking.

Among the more interesting anthropology projects have been those offered as UREP's SHARE projects, such as identifying public attitudes toward AIDS in London, Amsterdam, and Rio. The AIDS projects have been among the most heavily subscribed at UREP.

In Rio, volunteers were needed by researchers Cajetan Luna and Richard Brown to try to understand the social networks of the homeless children with AIDS who spread it through prostitution, as well as the attitudes of the public health officials in the country. Brazil has the second-largest number of reported AIDS cases in the hemisphere—more than 10,000 reported, and no one knows how many are unrecorded.

Going into the streets of Rio has its hazards, and the project leaders sent only those who were fluent in Spanish or Portuguese and who had the necessary courage. Volunteers who wanted to could hang out with the homeless children, trying to find out how they adapt to survive. Most of the childen are also involved with drugs.

"I took a backpack and went out to mingle," one volunteer reported. "I had no formal questions; I just talked to them. Then when I got back, I wrote down what I had heard." It wasn't pretty, he added.

Another volunteer, Andrew Forster, a medical student who speaks Portuguese, said that some street kids had no idea how AIDS is transmitted. "They'd heard of it, but it's the last thing they have to worry about. When they're on the street and they have to have sex to get money or food, this disease that will kill them 8 or 10 years down the line isn't that big a deal." He was impressed with the social support groups that were in place, such as a special school for street children and an orphanage for the homeless, as well as nonmedical support groups for those who are dying from AIDS. "But the problem is money. They only get a hundred doses of AZT a year, and that goes within a few days. Hospitals are overcrowded," Forster says.

The fundamental problem is that the seriousness of AIDS just has not gained a lot of attention. Anthropologists are busy trying to understand what attitudes toward sex create an environment that allows AIDS to flourish without any checks; and secular and religious support groups are trying to educate prostitutes and transvestites about AIDS. Forster says that part of the problem is that no one realized that some Brazilians' sexual identity is so fluid. "Many men are bisexual but don't consider themselves homosexuals if they are the dominant partner," he says. And heterosexual anal sex is often used as a method of birth control or as a way to maintain virginity. "But now that soap opera stars are starting to drop off with AIDS, that will have the same effect as when Rock Hudson died here. Now they'll start to realize it's not just gays who are getting it," Forster says.

Carl Hopkins, another volunteer who is a retired public health physician, interviewed government officials.

My assignment was to interview various government and private agency people who were involved in the AIDS epidemic, with the mission of finding out what they knew about it, what they were doing about it, and their attitude toward it. We were briefed before by two project leaders, a pediatrician, and a medical anthropologist, but our questions were not structured.

I had spent two summers on UREP projects interviewing health officials in London about their attitudes toward AIDS. Because my career was in public health before retiring, I felt that the AIDS epidemic is a major event of this century, and I wanted to be a part of it. I had done some homework, but the British experience was eye-opening because they were so much more concerned and better organized than we were.

In Rio, however, nobody was in charge. There was

chaos. Responsible people are shrugging off AIDS, saying, "We have all kinds of diseases; this is just another one."

But things happened as a result of the UREP project, Hopkins says.

> Several agencies and people who really cared but who were working alone and in the dark are getting together now. I had an invitation to a meeting of the Association of Prostitutes in Rio which is meeting with agencies and individuals relative to the problem. The project was exploratory; raising consciousness on AIDS was the most important aspect of it. I'm sure what little we did triggered some action.

In another part of Brazil, sharecroppers live with periodic famines, during which their children are the last to be fed. UREP has several projects with the goal of trying to understand why children are the first to suffer in an emergency. These range from observing parent-infant interactions to looking at the calories used in physical labor. The government recently instituted a school nutrition program for the children.

In a long-term study to determine the sharecroppers' energy inputs and outputs, researcher Alan Johnson of UCLA has been using UREP volunteers to interview sharecropper families in their homes. Johnson wants to compare current data with information he collected twenty years ago to see how people have changed as a result of social security benefits and other social changes. His project looks at what the sharecroppers eat, what they do during the day, and how much energy is used in work tasks, for example, cutting sugarcane.

Volunteer Robert Schilling describes how team members recorded the behavior of people at home:

> Each volunteer would go out with a Brazilian kid [who acted as interpreter] to visit homes at randomly selected hours. We'd walk in, and we'd ask, "Who is he? Where is this one? What relationship is she?" Then we would note whether they were cooking or grinding corn or taking care of the baby, and we would write it down. We kept a mammoth diary of every household by each hour.
>
> It's difficult and time-consuming; and you walk a lot. But it gives you the opportunity to know and understand people you would never meet under other circumstances. There is no other way you could live in a *fazenda*, this living arrangement sharecroppers have, which is run on a feudal system, and work with and begin to understand the lives of sharecroppers.

Geology

When the earth speaks, geologists listen. Like earthquakes, volcanoes have a language all their own that geologists are trying to decipher. What causes sudden eruptions or quakes? How far apart are they likely to be? What were the conditions in the past? Volcanoes are terrifying because they are sudden and unpredictable, yet, strangely enough, people who live near them tend to remain and after being displaced, return. Pompeii stands as a reminder of the speed of a volcano, as well as of the faith of those who moved back. Because volcanoes are so active and because everything they do is significant, they lend themselves to volunteer study. Scientists are monitoring many active volcanoes on earth, including Mt. Etna in Sicily, trying to discover patterns in their eruptions.

In Costa Rica, Smithsonian Institution volcanologist Bill Melson has been studying Volcano Arenal for the past 20 years. It blew catastrophically in 1968. Now, 21 years later, it still breathes fitfully: it can lie quiet for a few minutes or as long as 6 hours before blowing out lava and rocks. Melson expects it will wind down within 5 to 6 years. On previous expeditions, he has conducted excavations with volunteers to read prehistoric eruptions that appear to be spaced about 300 to 500 years apart. Soil develops when the volcano is quiet, he says; there is evidence of nine soils in 3,500 years.

What is it like to work with a volcano? Volunteer Sister Juliana Lucey, a research scientist, describes it:

> It's alive. It initiates events with an intricate plume. Then follows a resounding explosion. There could then occur whooshes, possibly rhythmic chugs and rock slides—huge rocks that break off the highly viscous magma and come crashing down the slopes. At this point, the volcano is sending out clouds of ash, which each of us hopes are directed to the field station where we are monitoring.

At a station based a safe two miles from the crater but still within the rim of cold black lava that looks like a huge vat

The staging area at the base of the active Arenal Volcano, where volunteers record its behavior. Photo: Sister Juliana Lucey, S.N.J.M.

of fudge, volunteers man a number of instruments. On a 24-hour-shift basis, they record the blasts of hot red lava spumes and time the length of eruption with a stopwatch. They also read the seismograph and estimate the intensity of the sound, which they compare with the sound-level meter. Another instrument measures the height of the blocks thrown out of the crater and fixes their location on the side of the mountain.

To chart the drift of volcanic ash, volunteers collect rainfall using rain gauges, and read windspeed indicators. Acid rain from the chemicals in the volcanic ash can kill vegetation on the downwind side.

Sister Juliana describes the rain forest: "Orchids grow wild; large, patterned moths park on our platforms; leaf-cutter ants work nonstop; iguanas cross our paths. In the very early morning, howler monkeys announce the dawn; squadrons of parrots glide by our station; tree frogs chirp untiringly; the numbers and kinds of birds seem to be unlimited."

She continues,

Monitoring the volcano is at times demanding; there are hikes to be made up steep slopes, and someone's sleep must be interrupted to be on duty. But there is another aspect to consider. Because the volcano is so active, the slopes undergo noticeable change.

To be on duty at Station B is a unique undertaking. 'B' is located far up the volcano slope and can be reached only after a strenuous hike up the rocky devastation zone. What a hike! But also what a view! You look out upon the rain forest and Lake Arenal. Also, you can hear and almost feel the chugging of the magma as it surges in the vent. I found myself dancing to its rhythm as I tried to gauge the wind direction. These experiences cause you to gain such a familiarity with the volcano that you personify it.

At night, the volcano was usually cloud covered, but our last night at Arenal was a clear one, and the volcano gave us a grand finale. The entire mountaintop shone red and gold; this was accompanied by the colorful movement of glowing rocks and gases and the resounding noise only a volcano can manage.

I am sure none of us will ever forget the experience.

Other geology expeditions in which volunteers can participate study wetlands, looking for indications that earlier global warming caused sea levels to rise. On these, volunteers take core samples for analysis of sediments. On other expeditions, geologists collect samples of the sediment layers that have been created in seasonal lakes, for example, in Australia, and these produce data that contribute to better understanding and management of groundwater resources.

Animal Behavior

Observing animals in the wild is not like watching them in a zoo. Volunteers who do it come back changed in mysterious ways, as if the experience awakens some paleolithic leftovers in the human psyche.

Basically, the work involves watching, being quiet, staying awake during long hours of watching, reading radio monitor signals, writing down every little movement, and weighing and measuring the animal if it is caught briefly, as well as describing every unique marking on it.

Just about any animal project is environmentally important: not only are habitats disturbed by man, for example, by denuding the forests, but most animals have been ruthlessly hunted at one time or another. Chimpanzees, for example, are hunted today because their biological and chemical similarities to man make them valuable in AIDS research. This trade is usually done on the black market, and very often, many chimpanzees in the group are killed, along with the dominant male, in the attempt to capture one or two alive.

TIMBER WOLVES

From Little Red Riding Hood to the boy raised by a pack of wolves in India, wolves inhabit our imaginations as creatures both dangerous and mysteriously caring.

If anything personifies the call of the wild, it is timber wolves. Adapted to the long harsh winters of the northern United States and Canada, wolves have been both revered as spirit guides by American Indians and savagely hunted for hides or because hunters saw them as competition for game. Wolves had no defense against guns.

Bill Robinson, Earthwatch project co-leader of a study of the habitat of timber wolves, grew up in northern Michigan where he remembered reading when he was in high school in the 1950s that there were 50 wolves left. "My friend and I just couldn't believe it because they were still bountying 25 wolves a year. Even though most of us were hunters, we thought it was a shame that we were losing them. And we did lose them."

In 1974, Robinson helped to relocate four wolves to northern Michigan to repopulate the area. "The results were that all four were dead within eight months, all at the hands of man: two shot, one trapped and shot, one hit by a car."

Shortly after, Robinson did a public attitude survey of people in the area to try to determine what they felt about wolves. Surprisingly, he found that hunters had generally more positive feelings than the public at large. "But within that group of hunters, there were some very anti-wolf sentiments."

Wolves inspire a variety of reactions, he has found. "Some feel sympathetic for the underdog, the misunderstood so-called villain; and they feel the wolf has been unfairly persecuted. Others feel admiration for this powerful predator; these are the same sort of people who are interested in great white sharks and have admiration for great strength. My own interest in the wolf is as a representative of the complete ecosystem and the symbol of the north woods and the wilderness.''

Timber wolves are gone in northern Michigan, but they still exist in northern Minnesota. Here the U.S. Fish and Wildlife Service has radio collared them, because they are an endangered species. If, for example, a farmer fears his cattle are being attacked by wolves, members of the U.S. Fish and Wildlife Service will relocate the animal to another place before the farmer can raise his gun. Robinson and Dan Groebner of the Wolf Institute use volunteers to monitor the wolves' territories and their response to humans, to see how much they use roads where they are likely to encounter people.

Most radio tracking is done by air, where they can get a fix once a day or once a week. On Robinson's teams, crews on 24- or 48-hour continuous shifts track them from a van over a network of roads.

Saurabh Misra, a volunteer who is a student at North Carolina State University, describes it:

We worked on 6-hour shifts around the clock, usually with two teams in two vans. When we got a signal, we would do triangulation, communicating with CBs between the two vans, to get a fix on the wolf in place and time. We can tell when they make kills, when they go back to their pups, when they're at the den. We found that wolves avoided humans everywhere except at a landfill area.

On our off-nights, the volunteers and researcher went wolf howling. Wolf howling sessions were done late at night, and we would walk into wolf territory in the pitch dark, where we had to be completely quiet. It was pretty scary. Then we would howl and listen quietly for a response. Once a pup came up close to us. Then we heard the mother growling nearby. But after two weeks of being around wolves, I don't think I'll ever be afraid of them.

Wolf howling does have scientific value, Robinson explains. A wolf has pups in late April, keeps them in the den for six weeks, and then moves to an open field in early June. "Then the adults hunt, usually singly in the summer, looking for animals like snowshoe hares, deer fawns, those that do not take a pack to subdue. To get an estimate of how many pups there are, we howl and count the number of voices that answer," he says.

Robinson remembers one summer when a French horn player from the Atlanta Symphony was a volunteer. "He plays the wolf in *Peter and the Wolf.* He did his wolf howl on the French horn and got a response. He comes back now on his own in the summer. He likes the wilderness."

SEA TURTLES

"It changed my goal for the rest of my life. I'll be a surrogate mother for the turtles as long as I possibly can."

—Volunteer on an FFR expedition to the Baja Peninsula

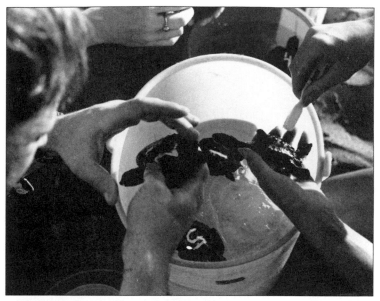

Volunteers handling baby sea turtles in Michoacán, Mexico. Photo: Walter Meyer

Leatherback sea turtles are endangered. Not only have they been hunted for their carapaces and their meat and killed on beaches where they come to lay eggs but they are particularly susceptible to environmental hazards, like plastic bags, which they mistake for food. No one knows much about turtle life underwater, but they lay their eggs in pits they dig in the sand on tropical beaches at night in the spring and early summer. One way to help turtles survive is to make sure their eggs are buried deep enough to protect them from high tide and poachers. Volunteers on several projects at different locations have an opportunity to do this.

Volunteer Jim Webber, a high school teacher who sponsors the expedition of a student who accompanies him each year, describes an Earthwatch project on the beach in St. Croix:

> We walked the beach from 8:00 p.m. to about 4:00 or 5:00 a.m. each night, single file behind the researcher and his assistants. They had designed a network in which we could patrol the whole beach. We used only moonlight and starlight because artificial light might alter the course of the turtles.
>
> When we encountered one, we would hear a kind of creaking and groaning. They're so huge when they move, and so quiet at the same time. They are just so impressive, it's amazing to be up against something that large, and touch its head.

Face to face with the focus of a Foundation for Field Research Project. Photo: Walter Meyer

When the turtle begins to lay eggs, she goes into a kind of trance. Then we could put on the lights, and everybody starts doing something: collecting data; collecting eggs, if necessary, to rebury; counting them; measuring the length and width of the carapace; noting any unusual marks. Then we attached straps underneath her, and using a pulley system, pulled her up to weigh her. Most weighed between 700 and 900 pounds. The researcher recognized some of the turtles from previous times. They had their eyes open for one they knew that was a 1,100-pounder.

Some turtles during the egg-laying season of March to July come up every 10 days. Others come up every 30 days; some every other year. Some researchers are investigating whether the turtles' pink eye spots, or pineal eye, are a kind of fingerprint.

The whole process takes about 1 to 1½ hours. Everybody worried about the dreaded dawn turtle, as they call her, the one that would come ashore when everyone was tired, and we'd have to stay with her until she had finished and was safely back in the water.

Their duty done, the mothers go back to the sea, while the eggs incubate in 3-foot pits in the sand for about 60 days when they all hatch at one time, and the hatchlings scramble up to the daylight. Then they struggle to the sea, where birds, man, dogs, and anything else that is hungry can interrupt the journey. Once in the sea, the hatchlings are floating snacks for any fish. Scientists estimate that only one percent of turtle eggs make it into healthy adult turtles.

Everybody loves the hatchlings. "They're like little

wind-up toys," said a volunteer. Some researchers put identifying numbers on their backs in the event one will resur-
face as an adult on the same beach.

Volunteers on turtle projects work all night, and they
work hard. The St. Croix project was limited to 10 days
because of its arduous schedule. Webber said, "After, we
would go back to the cottages, shower, eat something, and
try to sleep. But it was hard to sleep. In the afternoon we
would go play in the sea. Then we'd eat dinner and prepare
for another night's patrol."

On beaches near Savannah, Georgia, the loggerhead sea
turtle comes up to lay its eggs beginning in early May and
engages the attention of scientists and volunteers until the
hatchlings are safe in the sea in October. The Caretta Research Project (after *Caretta caretta*, the Latin name for the
loggerhead) has been in existence since 1972, in conjunc-
tion with the Savannah Science Museum at the Wassaw
National Wildlife Refuge. Spokesman Bobby Moulis, collections curator of the museum and a herpetologist, began as
a volunteer 17 years ago and has spent most of his life trying to save loggerheads. Their life is hard: the waterways
around Savannah are full of shrimp fishermen whose nets
trap the turtles and kill them. Part of the volunteers' duties
is to walk the beach and look for turtle carcasses.

But the biggest problem in the wildlife refuge is raccoons
and wild hogs, which feed on the eggs. Moulis has designed
4×4-foot screens, with 2×4-inch mesh, which allow the
baby turtles to get out and run for the sea but discourage
the groping arms of the raccoons. Night patrols by volunteers are the only foolproof way to guard the eggs, because

raccoons are frightened off by anything as big as a human, Moulis says. The hogs, too, are skittish around people, but one rout can devastate a nest.

It is impossible to predict which years might be especially fruitful. "Some are bumper years, some are down years," says Moulis. "You can't predict averages, because turtles tend to be on 2-year, 3-year, or 4-year rotation between nesting seasons." The 1987 drop-off, Moulis says, is probably due in part to the nesting rotations as well as to "the massive killout" in 1980 when 50 dead turtles washed up, some of them probably subadults who would have reproduced in 1987.

The Caretta project takes volunteers on one-week stays from May to September for $375, and $300 in September, which covers food, lodging, transportation at the site, and lectures. Work is arduous, involves walking 2 to 3 miles a night, enduring heat and humidity, limited electricity, rustic conditions, and no air conditioning.

ORANGUTANS
Physical anthropologists claim that orangutans are our close relatives. Orangutan researcher Birute Galdikas is trying to determine exactly what they are up to, using a band of orangutans formerly in captivity which she is rehabilitating back into the wild. On an orangutan reserve in Borneo, Earthwatch volunteer Will Hobbs reported what it is like to observe orangutans in their own habitat.

At 8:00 a.m., Tutut waded across the river and claimed the only dry spot of dry land around. So we spent the rest of the morning balancing precariously on tree roots above the muck, observing her.

By the time 2 o'clock rolled around, we had followed Tutut to a tree towering over a large muddy bog, which looks like Mother Nature's answer to toxic dumping. Like a good little scientist, I waded into the cesspool and began taking notes.

As I was busily taking notes, I suddenly felt something, a sensation, a tingling, a buzz, an irritation, and then it grew. It was a relentless itch, centered somewhere on my ankle. I scratched it with my other shoe, deep in the brown water. I rubbed it with a stick, a rock, a pencil, a tree, a root, a notebook, a backpack. The itch grew, until my entire leg was on fire.

But I held on and took notes, wanting so much to contribute to science, somehow. Then suddenly I felt a tap behind me.

I turned to meet another ex-captive orangutan, trying to get the food out of my coat pocket by circumventing the opening and chewing a hole in the bottom. Then, after becoming frustrated with my pocket, she began to shake me thoroughly. But like a dedicated scientist, I kept scribbling my notes.

Suddenly Tutut did something amazing.

She saw her nest on the other side of a bog 10 yards across. A dead tree stood on her side of the bog. She broke a large branch off the dead tree and positioned it at the edge of the bog, perpendicular to the tree. She pushed the dead tree down, across the branch, and it landed perfectly in a V formed by two trees on the other side.

Then I realized why she had placed the branch where she did. It braced her end of the log, which

An orangutan mother and her curious baby relax and play in Borneo. Photo: David Agee/Anthro-Photo

otherwise would have slipped into the bog. Tutut had constructed a bridge, across which she gracefully strolled with her children in tow.

I had just witnessed sophisticated tool use by a primate. Finally, I had made something to contribute to science. I was ecstatic.

Will Hobbs celebrated his sixteenth birthday a couple of days later. The above was published in Earthwatch's newsletter.

BLACK BEARS

An ancient Greek myth tells the story of Callisto, one of the goddess Artemis's followers, who violated her vow of chastity by getting pregnant. As punishment, she was turned into a black bear. Evelyn Angeletti, an Earthwatch volunteer on researcher Roger Powell's black bear project in the rugged forested mountains of North Carolina, tells the story of a black bear, which normally does not attack men, that was scaring hikers in the Smoky Mountains. She would rise up on the edge of the path, causing the terrified hikers to drop their sandwiches, which she would then eat. Then she would visit the camp farther up the trail, rise up on her hind legs, and scatter the campers long enough to have a little dinner. She became famous for her street smarts until a hunter got her, intelligence and all.

Powell does not know how many black bears there are in the Pigsah Bear Sanctuary, because the terrain is too rugged to make a systematic inventory. What is certain is that hunters and poachers have taken a heavy toll, acting out of long tradition of hunting bears and out of greed for the price that bear parts draw on the black market. Their gall bladders, for example, are used in some medicines.

Using volunteers, Powell attaches sardine cans to trees. The heavy smell draws hungry bears into traps. On regular early morning rounds, volunteers check the snares, and when they find a bear, race back to camp. Angeletti describes it: Powell gives the bear an injection with a jab stick that is a 12- to 14-foot metal pole with a needle at the end. When the bear goes down, everyone does something: one volunteer measures the respiratory rate and monitors the temperature to make sure the bear is okay; others keep a stat sheet, measure the overall length, and the length of the ear, the paw, and the tail, check the teeth, and put drops in the eyes and cover them with a bandanna so the creature won't be distressed. A trained person takes a blood sample, and if the bear is a yearling or older, it is fitted with a radio collar. It is a high-tech collar, which can be remotely triggered to release an anesthetic if Powell wants to check the bear at a later date.

Bears have no smell, says Angeletti. Their fur is rough. The smallest weigh 80 to 85 pounds; the largest, about 200 pounds, with the average closer to the low end of the hundred scale. "They're as large as an oversized German shepherd or a Great Dane," says Angeletti. Sometimes a researcher will remove a tooth from the captured bear because it carries a lot of dietary information.

The bears are more likely to fall into the traps in the spring when the lure of sardines is irresistible. Later in the season, the bears are well fed, and volunteers rarely see them. But they still have to check the snares each morning. "It was the worst and best part of it," says volunteer Eddie Bilezikian. "It rained a lot, and we got totally drenched. It was very arduous." At 5,000 feet, volunteers were happy to ride in the van, tracking the beeps of the radio collars. "The closer together the clicks, the faster the animal is moving. When the clicks slow down, we knew the animal was asleep. We calculated their locations to see how much of a range they needed, and, of course, the range they needed was much greater than the range they have after the encroachment of man," says Bilezikian. "These animals should live 30 to 40 years, but the lucky ones make it to 10. They leave the confines of the park and get shot."

Bilezikian and other volunteers collected blueberries and weighed them to check the quality of the berries that year. "The berry is the barometer of their food supply," said Bilezikian. "A substantial portion is so easy to get. The bears just take a branch of them, run it through their mouth and gulp them." Powell is also studying how much weight black bears put on before their winter slowdown and if females metabolize different from males. Angeletti says that the females' range seems to be related to food, while the males' range is related to available females. Powell believes that females need more calories during the winter because while they sleep, they give birth. Powell believes that to maintain their systems, they use about 4,000 calories a day.

Bilezikian and his wife, Ann, were not disappointed about not being able to see a bear. "If you really love these animals, you know they're better off without ever seeing you," Bilezikian says.

KANGAROOS

When Earthwatch opened its Australian office, Australian scientists submitted projects they thought would attract American volunteers. Among them was one investigating the habits of the Hastings River mouse. When Earthwatch public relations director Blue Magruder saw it, she recalls, "We said, 'Good grief!' and they said, 'But you said we should find appealing animals,' and we said, 'Yes, but the *Hastings River mouse?*' To Australians, kangaroos are pests. So the Australian volunteers go on the Hastings River mouse project and the Americans go on the kangaroo project."

But whether they are pests or large, furry, lovable animals, as Americans believe, kangaroos are being examined to determine if they do in fact bother ranchers' sheep as ranchers suspect. Researchers also want to know how kangaroo mothers treat their young and, in short, what kangaroos *do* all day.

Volunteer Delores Schmitz, a teacher, describes being around kangaroos:

We'd catch them at night by blinding them with a spotlight from the back of the pickup. Then several volunteers would run after the kangaroo and hold it while a researcher put a fast-acting tranquilizer in its mouth. Then we would cuddle it and put it in a bag usually used to carry sheep wool. Then while one volunteer held it—its eyes were still open, it was as if it was just dazed—we would measure the limbs and the overall length, check to see of there were any joeys in the pouch, then band it with a radio collar and set it free. The whole process took ten minutes. The hardest part was running along the rocky rutted road after them. That we left for the young volunteers, and they came back with some dandy bruises.

The other part of the project, said Schmitz, involved radio tracking the kangaroos from a tiny room atop a 45-foot tower.

Every fifteen minutes we would write down their behavior. We found that kangaroos were very territorial; they would move 1, 2, or 3 degrees in a 15-minute span of time. Many times they'd just sit and do nothing. We were ordinary people with no scientific background, so the researchers put a radio monitor on a tree to see if we were getting it right. And we were. They seem to be like our cows: they move freely but only within a certain territory.

We also monitored mama kangaroos from a smaller tower and wrote down their behavior every 15 minutes. The mama kangaroo seems to have three children at once: one out of the pouch hopping near to her; the other in the pouch but stretching out; and then in the bottom of the pouch, this tiny, tiny one, like a backup baby waiting for its turn to grow.

We also monitored a fox. One day the signal kept getting louder and louder, but we couldn't find it. When we found the fox, it was dead. That was so sad. You get so close to these animals when you're there because the scientists who are doing this research are so caring.

GAME PARKS

Alexander Peal, who left his position with the Museum of Man in San Diego to return to his native Liberia as Head of Wildlife and Parks, created the first game park in the country, called Sapo National Park. The first step was to hire the hunters as guides. Unlike hunters in this country, only certain people inherit the skill from their forebears in Liberia, and for Peal, that meant offering them a job with as much status. The second step was transecting the park into four sections of 3,000 meters for a cumulative total of about 60 kilometers to describe an area in which existing game could be observed, counted, and cataloged. For this, he has been using volunteers from the Foundation for Field Research. Access to the park is not easy. For the arduous trek, FFR hired porters to help carry the heavier gear.

Volunteer Howard Kaplan describes the first project to identify game. He is an executive with AT&T, newly retired.

It took us three days to get across the U.S., to Paris, to the Côte d'Ivoire, and then finally to Monrovia. We stayed in town the first night, with the Peace Corps the second night. Then we were driven in a station wagon with an escort and government officials to visit tribal chiefs to request permission to enter the park. The elders of the tribe came out, and we all sat under a tree in a circle, and they talked [in their Sapo language] and gave us permission.

We were deposited at the edge of the forest with the porters where we began to walk for five hours, crossing a river on a raft. The porters were noisy to keep away any hunters who might be there. We would walk for thirty minutes, then stop for ten. At one point someone yelled, ''Jump back!'' One of the women on our trip discovered she was a foot away from one of the poisonous snakes. If it gets you, it gets you for good in 3 to 5 minutes. She got out of its way. Sometimes tse-tse flies would come in patches. And we met soldier ants. You have to get your shoes and socks off fast and get them out because they sting.

We arrived at the campsite, which was a two-room building. We went for a bath in the river and looked up to see a python 6 feet away. Luckily, it was dead. It was 21 feet long. Unfortunately, the cots that had been sent by ship were not there because the paperwork had been lost. So we slept on the floor.

Our daily schedule began with breakfast, then we would go out on a trail and help build a treehouse to use as a blind from which to observe animals. We took turns sitting quietly there on hour shifts, taking pic-

A chimpanzee in Liberia holds a hammerstone in one hand and hard-shelled nuts in the other. Photo: Foundation for Field Research

tures, looking for animals. We saw duiker, which are little antelope, elephant tracks and some hippo tracks, wild boars, and red and black and white colobus monkeys. Every morning a golden cat would come to try to steal our chickens from us.

Richard Nisbett, one of the researchers, filed a report at the end of the season listing the game that had been sighted in Sapo National Park: bush baby, potto, lesser and greater spot-nosed guenons, the Diana and Campbell's monkeys, the sooty mangabey, red and black and white colobus monkeys, and the chimpanzee.

Chimpanzees are not easy to see in the jungle. Nevertheless, researcher Alison Hannah hopes to document evidence that chimps are not only avid tool users but possibly toolmakers as well. FFR volunteers will search for chimp tracks and spoor in the jungle and look for areas where chimps have used rocks and a flat surface on which to crack open nuts. Hannah wants volunteers to stake out known areas where chimps have used tools to see what their behavior is and to try to find new areas.

Long the victims of hunters, and today the objects of bounty hunters for use in AIDS research, chimps are skittish near man, and the jungle is thick enough to hide them. One of the goals of the FFR projects is to employ some of the bounty hunters as helpers to redirect their energies from hunting chimps to helping them.

Sea Research

Most of our planet is water, but the oceans remain mysterious places. Only recently have entire new species been found growing and thriving near hot air vents in the deepest parts of the ocean. Currently, climatologists are beginning to realize that the ocean influences our weather in ways previously unknown, as, for example, when dead algae rises from the bottom to the surface of the sea and significantly changes surface temperature.

What has been clear, however, for the past couple of decades is that the human population, through abusive fishing practices, has had a direct effect not only on mammals and fish that live in the sea but also on the coral reefs that grow near shore. Coral polyps are tiny creatures that leave skeletons when they die, and the skeletons form the coral reef. A living reef is an ever-growing reef. Coral reefs function like underwater cities with fish and mollusks living in ecological harmony. Here, sponges grow in crazy shapes and colors, octopuses keep hidden nests, and little fish find protection from predators by disappearing into clam mantles.

But overall, coral reefs are delicate places. Marine biologists are trying to find out just how delicate they are. When careless sailors drop anchor and hook onto a living reef, the reef will probably break and die. Chemicals from cleaning liquids and oil and gasoline from boat motors all pollute the living undersea environment. In some parts of the world, voracious fishermen have dynamited reefs to release mollusks and reveal octopuses. Scientists are trying to gauge how fast the reef is dying as a result of all these things; this is an area where volunteers can help by mapping and measuring reefs.

On some marine biology projects, nondivers man shore stations, where they record locations of various underwater features with a surveying instrument from information relayed by divers. Other shore projects use volunteers who walk the reef, looking for evidence of damage near the shore or interviewing fishermen about their fishing techniques and habits. Some projects involve sitting for a long time with a net waiting to trap certain kinds of fish for transport to an aquarium.

CEDAM

CEDAM, which stands for Conservation, Education, Diving, Archaeology, and Museums, is a good description of the organization's purpose. Begun twenty-three years ago by Paul Bush, a Texas businessman/explorer who brought scuba back from Europe (where it was invented during World War II by a team that included Jacques Cousteau), CEDAM was incorporated in El Paso. In its early years, it sponsored marine archaeology projects in Mexico. But for the past ten years, under the aegis of Rick and Susan Sammon, CEDAM has been involved in international marine conservation efforts.

Using paying volunteers, CEDAM sponsors projects in various parts of the world which help clean up and map reefs, survey marine life in a particular area, and collect fish for aquariums. On recent projects, volunteers collected live fish in the Red Sea for the New York Aquarium, put mooring buoys in place in an effort to save the reef in Belize, did cleanup dives in the habitat of the endangered manatee in Crystal River, Florida, and did a marine life survey project in the Galápagos.

One of their more successful projects has been to develop guidebooks specifying the rules and regulations for marine parks in the Seychelles, Belize, and off the Kenya coast. Less developed countries lack time, money, and labor to be involved in conservation projects, Susan Sammon notes. Volunteers have helped write and produce the guidebooks, which have been donated back to the parks, which then sell them to raise revenue for their maintenance.

"We need to focus more awareness on the fact that if we don't start taking care of marine areas and establishing sanctuaries, we won't have them anymore," says Sammon. One of CEDAM's strong points is its ability to energize public concern about the waterways. In 1989, it asked scholars, scientists, and marine supporters from the United Nations, the National Geographic Socicty, and private foundations to identify "Seven Underwater Wonders of the World,"

CEDAM diver meets sea turtle in the Galápagos Islands. Photo: Rick Sammon/CEDAM

which were endangered. Coming together in Washington, D.C., the committee, which also included actor/conservationist Lloyd Bridges and astronaut/aquanaut Scott Carpenter, identified the most troubled areas as the Republic of Palau in the South Pacific, Lake Baikal in the Soviet Union (so big and so deep, says Rick Sammon, that "if you took all the waters in the Great Lakes and put them in Lake Baikal, it still wouldn't be full"; now threatened by industrial pollution), Ras Muhammad and the North Red Sea, the Galápagos Islands, the Belize Barrier Reef, the North Great Barrier Reef in Australia, and the Deep Ocean Vents. President Corazon Aquino of the Philippines, whose coral reefs have been seriously damaged by cyanide and dynamite fishing, applauded the "Seven Wonders" project and wrote that she hoped it would "instill in all of us, regardless of creed or theology, the need to protect and preserve our fragile marine ecosystem."

CEDAM proposes to produce a documentary on all the sites, a traveling photography exhibit, a book, and a behind-the-scenes video.

For volunteers, the cost of one- and two-week projects ranges from $1,100 to $3,500, which includes round-trip airfare from Miami, lodging, meals, diving, and lectures. CEDAM membership includes an initiation fee ($7 for students under 25; $20 for others) and annual dues ($8 for students; $20 for others). Dues, fees, and contributions are tax deductible. There is no upper age limit; children are welcome with their parents but as nondivers.

THE COUSTEAU SOCIETY

"People protect what they love," says Sandy Bond, spokeswoman for the Cousteau Society in Norfolk, Virginia. "Our

155

purpose is to protect and preserve the quality of life for present and future generations. When people see what needs protecting firsthand, they invest something of themselves in it."

Jacques Cousteau epitomizes the caring scientist whose courage matches his curiosity. Aware that his kind of enthusiasm is contagious, the Cousteau Society sponsors two expeditions a year for paying volunteers. These trips offer an intensive program of lectures by a marine biologist and actual participation in some aspects of marine repair.

"It's very fast-paced both intellectually and physically," says Bond. Participants make two dives in the morning, and in the afternoon they can go to a lecture and then go back to work in the water. For eight years the society sponsored a program in the British Virgin Islands in which participants mapped and measured the coral reef so that scientists can monitor change to find out what effects nearby development have had on the reef.

"We get a hundred letters a day from people wanting to be on the *Calypso*," says Bond. "That's not possible. But these expeditions give an idea of what it's like to be a marine biologist and a taste of what it's like to be in the field." Bond says a lot of students go on Cousteau expeditions to find out if they want to study marine biology. "Our staff marine biologist is a dynamic teacher, so students really get a feeling for it. Some say, 'This is great!' Others say, 'This is too hard!' "

The Cousteau Society has been sponsoring programs for about eighteen years and takes about thirty-five people each expedition, usually offering two expeditions a year. The teachers, lawyers, film producers, and business people who go on them usually get fired up about marine conservation, says Bond. After, they share what they have learned with their communities, hoping to inspire that love that will instigate protection. "People make it their mission to protect the resources of the planet," Bond says.

The cost for two weeks on a project in Fiji in 1990 was $3,650, which included round-trip airfare from Los Angeles, full room and board in a resort hotel, and all program activities. Jean-Michel Cousteau participates in all projects. Diving is not a prerequisite, but it is recommended "in order to participate fully." Minimum age is 16.

HUMPBACK WHALES IN HAWAII

Whales began on land millions of years ago and then moved into the sea. Their skeletons still have little hind legs and a tail, and whale babies look like pigs, before the nos-

Whale researcher preparing for a underwater videotaping dive. Photo: Stephanie Ocko

trils develop into the blowhole on top of their head. Scientists are only beginning to understand them. Whales, said a researcher, are sexy and mysterious. Studying them is hard: they are too fast and streamlined for radio transmitters, and some whales can dive to depths of a mile or more.

Above the Equator, humpbacks migrate north in the spring, gathering off the coasts of the islands of Hawaii with their newborn calves to mate, then continuing up to the Alaskan coasts, where they feed on plankton and fish. Humpbacks are the most athletic of whales; occasionally they jump out of the water, head first, spin in midair, then land on their backs. At other times, they slap the surface of the water with their tails, called flukes, roll over like barrels, or simply lie on their backs on top of the water.

The horizon from the shore station on the Big Island of Hawaii is as clear and clean as the smooth swath of inky blue Pacific. Haze from the volcanic activity a couple of hundred miles south hangs like pink gauze over the bay, but at the top, breaking through like the castle in the *Wizard of Oz*, is the snowcapped peak of Mauna Kea.

Suddenly the glassy surface of the water is broken by the splash of foam and a flash of black and white: a whale surfacing to blow before diving, tail up.

"Whale fix, 508," Rick says from the theodolite.

"Note," says Adam from his folding chair, getting the whale in his binoculars.

"Note 105," says Susan, punching it into the computer on the chair in front of her.

"Note 105," says Neuille, on the other computer.

"392, 292, one's inshore," says Adam, the project leader on this Earthwatch expedition. "What's the azimuth?"

The theodolite fix will later be correlated with observed recorded whale behavior. Before long, the sea seems to be teeming with undersurface activity. Everywhere you look are whales, many in pods of two, swimming lazily south to north or north to south without any apparent connection or reaction to one another. It's like a Sunday stroll. What exactly are they doing?

"We don't know," says Adam. "They don't eat when they come here. They come only to mate. The only way you can really tell the sex of a whale is from underneath, and it's important to be able to know if they travel in male-female groups, or all-male, or all-female. Many of the mothers have newborn calves with them. Often they rise out of the water or roll over and slap the surface with their flippers. Yesterday I saw two whales side by side, one with its head up, the other with its tail up, rise straight out of the water," Adam says.

"A calf came within three feet of my face in the Zodiac," says Suzanne, one of Adam's assistants. "I don't think he knew what he was doing. I was looking in the water when he jumped up and almost capsized the boat."

Do they sing?

"Oh, yeah," says Adam. "You can hear them if they are close to the boat."

"They say if they're singing while you're swimming, you can feel it in your chest," says Rick.

One of Adam Frankel's primary research objectives is to understand what whales say to each other by playing back recorded song and observing their behavior. The incident in 1985, when a humpback got lost in the Sacramento River near San Francisco, illustrated the effectiveness of recorded song. When Frankel heard about the scientists' frustrated

159

attempts to turn the whale around and head it out to sea, he sent them a copy of a recording of whale feeding sounds as a lure for the hungry whale. In two days, the whale turned around and swam down the river and to freedom in the ocean.

Because of their predictable migration routes, humpbacks were ritually slaughtered by whalers for the better part of a century and a half. Although the species has been protected by the International Whaling Commission for several years, their numbers are low and their reproduction rates are weak. The whales are hard to study: they can't be studied like birds, which have nests, or elephants, which leave spoor. Their social structure and needs baffle scientists. Understanding them, Adam says, "means we are in an improved position to protect those needs."

With assistant Christine Gabriel at the wheel, the 17-foot Boston Whaler skims across the surface of the sea, running parallel to the northwest coast, about a half-mile away. The motor grinds and moans as the boat slaps the water, and the wind plays with the ever-present smell of gasoline. In the bow, Neuille sits with a record book; holding a camera with a telephoto lens, assistant Mike Hofhines stands and waits.

On the port side, two whales surface and dive. "Half a fluke!" says Christine, noting that the whale probably lost it in a fight with another whale or to a predator, a shark or a killer whale. It looks as if something took it away with a big bite. They surface again and engage in what looks like play. "We don't know whether they're playing or not," she says.

"Head rise, couple of blows, rolling pec slap, flukes up,

breach," Mike calls out to Neuille who records the behavior. The pair dive and do not reappear for another five minutes, when they are half a mile away.

On the starboard side, a mother and a calf ride in tandem. The mother is unbelievably big. "She's just had a baby," whispers Christine. "She's still fat." Suddenly, she points to a shadow underwater a few feet away, between us and the mother and baby. It is policy not only in the boat but also on land to be as quiet as possible, because whales have been proven to have fantastic hearing.

"Watch out," says Mike softly. "There might be a big splash."

Before he can finish, a bull whale surfaces, as large as the mother but with a nose that is covered with huge bumps, made white from fighting. With a quiet ferocity, he eyes the boat, rides as big as a submarine next to the tiny Whaler, then breathes out with a kind of indignation, and dives. His message is clear: the pair on the other side of him were not to be toyed with. "Whales are not monogamous," Mike says. "The bull whale is called an escort." Later, Adam says, "He could care less about the baby; he was protecting the female."

Mike takes the video camera in its underwater casing out of the box, puts on flippers and mask, and prepares to get some pictures to try to sex the pair, one with half a fluke, that have reappeared on the port side. Christine maneuvers the boat ahead of the pair as Mike slips off into the midnight blue water and the whale pair surface and blow. But, in the water, Mike loses them after they dive. The ocean is deep.

As Mike takes the wheel, Christine attaches a microphone to an amplifier and slips it into the water. The air is suddenly filled with whale song. "That's a singer," she says. "Singers" are always male, and they sing usually in a vertical position, nose down, probably to attract a mate. At first, it sounds like a dialogue, with a mimicked phrase in a different tone, then it falls to a deep sound and rises to a squeak, then mimics what sounds like a motorboat. Whale song is complex, with rhyming patterns and repetitions that go on for hours. On the boat, Christine turns down the sound. She shakes her head almost apologetically and whispers, "He's really singing *loud*!"

Christine leans over and scoops something out of the water. It is gray and about the size of a quarter. "Whale-skin," she says. "It carries all kinds of information." she carefully puts it into a bucket of water to save for later analysis. Every little bit of information counts with whales.

The cost of this two-week project including food and lodging is $1,850. Volunteers share a house with the scientific staff and take turns cooking.

HUMPBACK WHALES IN ALASKA (INTERSEA)

Off southeast Alaska, in the long days of summer, humpback whales gather to feed. Humpbacks are baleen whales and eat fish and plankton. "In pods of three to twenty-two," says Russel Nilson of Intersea, a private whale research group based in Washington, "the whales target schools of herring. Then they dive and sing a song, a very loud vocalization that lasts over a minute, which stuns the school of herring. The herring ball up, and the whales rush up from the depths with their mouths open in an intricately choreographed maneuver where each whale is in the same position relative to another every time they do it."

This amazing behavioral activity was discovered by Nilson and Cynthia D'Vincent who have been researching humpback feeding behavior for the past eight years. They were the first to record the fact that baleen whales actually cooperate when they come together to feed. Their work has been the subject of a *Wild Kingdom* film in fall 1989 and a *National Geographic* magazine article in 1984. *Voyaging with the Whales*, by Cynthia D'Vincent (Toronto: Oakwell Boulton Publishers, 1989), describes their research and contains some astounding photographs, taken by D'Vincent.

Volunteers aboard the *Acania*, a 126-foot former government research vessel (and before that, in the glitzy 1930s, actress Constance Bennett's private yacht), attend onboard lectures with a scientist and can choose their own out-of-water research activities, from collecting oceanographic data and running a computer program in which individual whales, identified by their flukes, are logged and tracked each summer to photographing their behavior and researching how the humpbacks interact with each other and with the Alaskan environment. The waters teem with whales and plankton, and D'Vincent says that the feeding chorus line is "the most spectacular behavior of the entire animal kingdom on earth."

Eleven days aboard the *Acania* costs $2,490; that includes not only comfortable berths but also food cooked by an onboard Seattle chef. There are no age limits on either end; D'Vincent's kids often go along on the trip.

Architecture: Arcosanti

Arcosanti is an experimental city struggling to be born in the desert of central Arizona. But unlike other cities that carry the centuries of their history in numerous structures on numerous streets, Arcosanti will be a huge single cluster of units containing living space, commercial centers, and cultural arenas for 6,000 people. Two giant greenhouses on either side will grow food for the population and provide winter heat; and the total absence of cars within the city (cars for use outside of Arcosanti will be parked in a garage belowground) will encourage a newfound human energy and interaction.

Or so believes its inventor, Italian arthitect Paolo Soleri. Once a student of Frank Lloyd Wright, Soleri conceived of Arcosanti in the late 1960s as a solution to the problems not only of environmental abuse—air pollution, inefficient waste disposal, chemical pollution of waterways, overuse of water systems, foodstuffs tainted with herbicides and pesticides—but also of personal isolation which are created by living in single-family homes in suburbs, watching television as a pastime, and driving in cars, all of which minimize the possibilities of human interaction.

Soleri calls it *arcology*, a combination of architecture and ecology. Linked to a harmony with the environment in philosophical ways, Arcosanti also relates in real ways, by using the earth as forms in which to lay concrete or as material for tiles and windbells and by using wind and sun as energy sources. Symbolically, the city will rise out of the earth, and Soleri envisions a day when Arcosanti will be a city for a million people.

But after the activity that the initial construction of

One of the guest rooms at Arcosanti. Photo: Tomiaki Tamura

Arcosanti generated among idealists in the early 1970s, Soleri found it harder and harder to get funding. Critics complained that his architecture was brutal, idealistic, radical, rigid, fanciful, impossible. But few ideas of genius fly the first time around; for Soleri, the stops and starts have provided the challenge to go on. In 1990, Arcosanti received a grant to start construction on the first section of one of the greenhouses.

At the moment, Arcosanti is an unusual cluster of pre-formed concrete boxes with big round windows, which house offices and apartments for the 60 to 70 people who live and work there, and open ''apses'' or shell-shaped spaces with aprons of stepped concrete that function as areas for the kilns where bronze wind bells are made and as places for concerts. Arcosanti sponsors a "Meeting of the Minds" seminar week twice a year when philosophers, writers, psychologists, musicians, and scholars get together informally to share ideas and talents. About 50,000 people from all over the world visit annually.

New goals target finishing one of the greenouses, building a hotel within the next three years, and providing living space for 500 people, what Soleri calls the "critical mass." Anyone who is interested in seeing how a city can actually be developed from an idea, into blueprints, then set into the tangible struts and walls that occupy a space should go to Arcosanti. Anyone who is interested in seeing how a city that is tied to the earth and ecologically sustained and environmentally nonabusive is welcome to spend as many as 4 weeks there. Living space is spartan but architecturally interesting—slanted walls and round windows, unexpected turnings of stairs, surprises like tiles on the walls and wind chimes everywhere, and places where there is only the wind.

Some of the food is grown at the site; other food is bought at co-ops outside Arcosanti; it is cooked and served in the cafe. "People sometimes ask us if we make our own clothes," says spokesman Scott Davis, an urban planner who is Soleri's assistant. "We are not self-sufficient; we're self-*reliant*. We still use the outside world. It's important to make that distinction," he says.

Arcosanti is amazing in a variety of ways. It is a visionary city coming into architectural reality, and it is also starting from scratch ecologically, so it can avoid all of the urban pollution problems that have developed over the past centuries. Foreign visitors are particularly interested in it,

View of Arcosanti in the Arizona desert. Photo: Tomiaki Tamura

because few other countries have the space in which to build a city from the ground up. All visitors come away with the sense that, despite what critics say, the only way to know if something completely different might work is to try it.

A 7-day seminar costs $400 and includes lectures, tours, informal conversation with Soleri, plus room, board, and a chance to work on whatever project is going on at that time, which might be construction. If you choose, you can stay on for as many as 4 weeks at a cost of about $40 a week for food and board and the chance to participate in the construction. No prior experience is required; and the $40 is *not* a misprint.

Elderhostel also offers a one-week seminar at Arcosanti; at the end of the week, Elderhostelers make clay bells or clay tiles that will be put on the walls as decoration. The extraordinary bronze bells that are made and sold at Arcosanti contribute to a good portion of its annual income.

The Human Environment

"We were willing to work where no one else would."

—Guy Bevil, founder of
Amigos de las Americas

"People should realize that that's the real opportunity: to get to a place and to do things that they would never be able to do otherwise."

—Veteran volunteer

Writing in a recent issue of *World Watch*, Alan Durning, Worldwatch Institute senior researcher, defined poverty in the world in 1990:

> In 1978, Robert McNamara, then-president of the World Bank, gave what stands as the classic description of absolute poverty: 'A condition of life so limited by malnutrition, illiteracy, disease, squalid surroundings, high infant mortality, and low life expectancy as to be beneath any reasonable definition of human decency.'
>
> As McNamara's words suggest, poverty is far more than an economic condition. Although usually measured in terms of income, poverty's true horror extends into all aspects of an individual's life. Susceptibility to disease, limited access to most types of services and information, lack of control over resources, subordination to higher social and economic classes, and utter insecurity in the face of changing circumstances are the norm for a poor person. Flowing from these dimensions, poverty's psychological toll is equally severe: the erosion of human dignity and self-respect.

Durning goes on to identify the poor: four-fifths of the poor live in rural areas; within the poor, the poorest are defined as those "who spend 80 percent of their income on food but still lack sufficient calories to meet their metabolic needs"; they tend to be illiterate, female, and landless, yet they work hard. Durning quotes Robert Chambers of the Institute for Development Studies in Brighton, England: "People so close to the edge cannot afford laziness or stupidity. They have to work, and work hard, whenever and however they can. Many of the lazy and stupid poor are dead."

In many countries, grass-roots organizations are mobilizing the poor to at least begin to "break out of the poverty trap." Durning goes on to define what components are essential: "female education, redistribution of farmland, empowerment of communities to control local natural resources, and extension of credit, clean water supplies, primary health care, and family planning services."

It has been said that *any* help the First World gives to the Third World is important in the changes it can bring about in another person's life. But very often it is the *individual* contact that is more effective as well as more reliable than financial aid poured from one country's banks into another's. Several opportunities exist to help on a small scale; some are reported below, and others are listed at the end of this book.

TEACHING OPPORTUNITIES

In a tiny, dimly lit, windowless schoolroom in a remote part of Tanzania, under a tin roof, on dirt floors, the children carefully form their letters on tiny slates held in their laps. The teacher has the only book in the classroom. Yet the feeling among the students is one of rapt attention, enthusiasm, and excitement. When one UREP volunteer returned from a project in which she was allowed to lecture in a classroom, she was so moved that she sent out a request to her entire school district to collect unused textbooks. She then boxed them and sent them to the school in Tanzania, because she knew every word would be valuable there.

"Teachers take a lot for granted in this country," said UREP's director Colvin, "and they have problems and complain a lot. But when they see how schools operate in other countries, it opens their eyes." Another teacher created a project for her elementary school children in which each researched a native American species of plant or animal, drew it, and wrote a description of it. Then the class bound the pages into a book, autographed it, attached their pictures, and sent it to a school in Kenya.

In the belief that teachers have a "multiplier effect," UREP has a special scholarship program for teachers to join in projects lasting two to four weeks in the summer. Under a grant from the National Science Foundation, the program gives teachers a prefield orientation to prepare them to develop a significant curriculum project on an aspect of the field project. A follow-up workshop after the team returns to the United States organizes the experience. "Brainstorming with other teachers was the extra push I needed," wrote Iris Weiner in the UREP newsletter. "I would demystify science and use my adventures to explain the scientific process. I wanted to show my students how exciting science could be."

In 1990, under a new liaison with UNICEF, UREP-sponsored teachers worked on an environmental education workshop in Ecuador with local teachers and scientists from the University of California and Ecuador. ''The NSF teacher grants intend to establish sharing of cultural classroom ideas to inculcate a global perspective,'' says Colvin.

Earthwatch also has a teacher scholarship program. Celeste Gauthier, a teacher of deaf children, described her experience: ''As a teacher I was excited about the prospect of sharing with my students a firsthand account of how people in another part of the world live. I had no idea that I would come away with such a profound admiration and love for the Maori people. Geography is so much more than just labeling areas of a map. If I accomplish only one thing this school year, I will be satisfied if it is this: to teach my students that every speck they see on a map is important. Things are happening there.''

For teacher programs, contact:
UREP
Teacher Program, Desk DO2
University of California
Berkeley, CA 94720
(415) 642-6586

Earthwatch Education Office
Earthwatch
P.O. Box 403
Watertown, MA 02272
(617) 926-8200

WORLDTEACH
China, Costa Rica, Kenya, Poland, and Thailand were the countries that requested teachers of English from World-Teach, a nonprofit social service program, in 1990-91. WorldTeach recruits paying volunteers who are willing to commit a year to teaching in a foreign country, living in housing provided by the host country on a stipend that is equal to a teacher's salary in the host country. Volunteers go to the country in a group and can expect to teach at the high school or college level. The only requirement is a bachelor's degree and a one-semester course in teaching English as a Foreign Language, or 25 hours of experience in teaching English as a foreign language; learning the host country language is not a prerequisite.

The average age tends to be in the mid-twenties, says a WorldTeach spokesman, but there is no upper age limit, and retirees and those at mid-career looking for a fresh outlook are encouraged to apply. The experience is one of total saturation in the host country culture, with the opportunity to travel during school holidays.

Fees range from $2,990 to $3,750, which covers round-trip airfare, health insurance, and administrative costs. World-Teach will give help in fund raising ideas; student loan payments can be deferred for the year spent teaching. Groups leave in February, May, August, October, and December.

HABITAT FOR HUMANITY GLOBAL VILLAGE
WORK CAMPS
At Habitat for Humanity International's Global Village Work Camps, volunteers with no previous construction experi-

Coahoma, Mississippi: a college student on spring break helps children build a new house for their family. Photo: Habitat for Humanity International

ence actually participate in the building of a house. There is nothing complicated or fancy about the house, but in many parts of the world, simply owning a home catapults the owner into a new level of self-respect and the family into a new integrity.

Volunteer donations make it possible for Habitat to offer no-interest, long-term loans to the buyer who works alongside volunteers in what's called "sweat equity." Families apply for Habitat mortgages and are chosen on the basis of special needs as well as the ability to make payments, which are then fed back to Habitat to finance more houses. The willingness to work in partnership is equally important. "Sweat equity increases the sense of home ownership," said a volunteer.

On a project in Guatemala, a Habitat volunteer helped carry bricks from the factory to the house site. "Bricks" in this case are cinder blocks, the building blocks of the Third World. Because the house was in the city and the owner had a full-time day job, it was being built by contractors. But the owner put in his "sweat equity" after hours at the factory making the bricks for his house.

On the Guatemalan project, volunteers spent a week building and a week traveling in Guatemala looking at previously constructed Habitat houses. In one case when they encountered owners living with a pool of stagnant water from poor drainage in the front yard, they pitched in and dug the proper canals. Volunteers did tasks requiring special skills, such as mixing the mortar and laying the bricks, under the tutelage of the local builders. "They were hesitant at first to let us do it," said a volunteer. "They were accepting and friendly, but they weren't sure what to do with us. Then they taught us how, using Guatemalan tools, and it worked out well. By the end of the week, we were really communicating. I think they were pleased to have us there," he said.

With 61 work camps in 30 countries, Habitat offers the opportunity to get to know another country in two weeks in a way that a regular traveler could not. Lodging is often in a house like the one they participate in building, without electricity, with a latrine, and showers "in which you pour water over yourself from a bucket." Each volunteer shops for food in the local food market.

All overseas Habitat groups require at least one person who is fluent in the language, but otherwise, language facility is not a requirement. The only requirements are good health and a minimum age of 18.

In the host country, each volunteer is responsible for the costs of his or her own food, lodging, transportation, and insurance. Participants must contribute an amount to equal the cost of one new house in the country. Houses usually cost between $1,000 and $3,000. "In Nicaragua, for example," explains coordinator Margaret Harvey, "an average house costs $1,000 to build. That means that each member of a group of ten will pay $100. In some cases, where special interest groups go on a project, they will hold a fund raiser before to raise the cost of the house." Habitat fees, a cook, and the use of a truck are $400.

Habitat is a Christian organization and will place a Bible in the house they have helped to build; but the organization is ecumenical and often international in its group formation, and all are encouraged to join. Habitat tripled its growth in 1990, from 300 to 750 participants. The median

A joyful woman leads a traditional rite of celebration for her new house in Zaire. Photo: Habitat for Humanity International

falls strongly among women 21 to 27 years old "doing something after college and before a career," says Harvey. "But an equal number fall in the 49- to 59-year-old group of both sexes."

Habitat suggests taking at least six months to prepare for a work camp. Detailed information includes usable phrases and a vocabulary in the host language as well as the country's cultural expressions. "We can't emphasize enough how you need to be culturally aware and sensitive *before* you go," says Harvey.

Global Village Work Camp Coordinator
Habitat for Humanity
Habitat and Church Streets
Americus, GA 31709
(912) 924-6935

HABITAT FOR HUMANITY IN THE UNITED STATES
There are eleven jobs for the 351 people who live in Coahoma, Mississippi, and the average income is $4,000 a year. Residents live in shacks with outhouses, without running water. Under a project called Operation Springbreak, Habitat for Humanity has made it possible for students to spend any part of 6 weeks helping to build houses in Coahoma. Last year, 450 students from 32 schools and colleges in the United States raised money and brought their labor to help build six houses.

"I always thought Mississippi mud was a dessert," says Kim Budke, a Coahoma volunteer who now works at Habitat. She helped construct one of the wood-frame houses with painted siding and shingled roofs, with interior plumbing and electricity. "It really made an impact on me personally," she said. Seeing children growing up without anything opened her eyes. "When you're a child, and you've always lived in a shack, trying to keep warm and dry, sharing a bed with three or four brothers and sisters, having a house means those needs are taken care of. That allows kids to make school their primary focus and parents to focus on other things like education and finding a job, things we take for granted."

But to build a house in Coahoma costs $10,500, which because of the lack of jobs, owners pay back with welfare payments. This does not leave much for other things. Contributions from students are considered "seed money," which, along with the mortgage payments to Habitat, is recycled back into the community to build another home. But when the money is not there, says Budke, and the payments aren't made, construction must stop. An international group called World Vision is assisting in economic development in the area.

For Budke and for many students, the experience at Coahoma challenges traditional concepts of charity. It is difficult to accept the limitations of what you can do. "When we went in," Budke says, "we were very conscious of not saying, 'We've got all the answers; this is how you do it.' The community was very open and wonderful, and we emphasized our equal partnership with the families in the building of the houses."

But going back later and seeing screen doors with unrepaired holes and the front yard untended brought her closer to a real understanding of poverty. "Providing a house doesn't solve all their problems. How do you expect somebody to take care of something if they've never had anything to take care of before?" she asks.

For volunteers trying to close the gap between their ideals and reality, Coahoma and many other Habitat sites in the United States can provide a living workshop. Building a house is a small step in the enormous reconstruction after the devastation caused by poverty. "You can take off your rose-colored glasses even when it's very easy to get cynical about it. It was affirming for me to see that we *can* have these ideals, and some of them might actually make some impact, even though it might be smaller than we had hoped," says Budke. "The problems out there are really big. But for the people in the community it meant seeing somebody come in, having made a promise, and following through on that promise. *That's* important."

Habitat offers several other work camp projects in the United States; and projects like Coahoma have been expanded to five other sites across the South during 1991.

AMIGOS DE LAS AMERICAS

"The most rewarding experience I had was knowing that some of those 4- and 5-year old smiling faces would still be around to be 17- and 18-year-old smiling faces, because of me and Amigos."

—Volunteer

On vacation in Honduras in 1964, a young man from Houston traveled into the outback and saw something he had never seen before: children dying of polio and measles, diseases that kids in this country are routinely vaccinated against. He asked the mayor why his village had no vaccines. "This," said the mayor, "is the village that even God forgot."

The tiny village in Honduras became the first to receive not only vaccines but a group of twenty-five enthusiastic young people to administer them. The five provinces where they inoculated children with the Salk vaccine in 1965

Protecting the water supply of the remote villages is the chief aim of the community sanitation/latrine construction projects of Amigos de las Americas. Photo: Amigos de las Americas

reported only eight cases of polio during a later epidemic.

After his vacation, seeing where he could make a real difference, Guy Bevil went back to the Houston church in which he was the head of a youth project and took care of the details that would make it happen. Soon another church in Houston joined them, and before long they became known nationally, so that ten years later they had 300 volunteers from 23 states giving inoculations to children whom God had no longer forgotten.

Today, Amigos de las Americas is a nonprofit, nonsectarian organization in Houston which trains and sends to the field 500 to 600 volunteers for four to eight weeks at their own expense, usually between $2,000 and $3,000. Most volunteers are enlisted on the community sanitation program being run in Central and South America, in which volunteers teach people in remote rural villages basic health measures ranging from covering food to where to build latrines to protect their water supply. Volunteers build on the average of 2,000 to 2,500 latrines a year. As one volunteer commented, the experience "makes you appreciate the small things like lights, bed, toilets. Before, poverty to me was just a word."

In Ecuador, volunteers have vaccinated more than 50,000 animals against rabies; they reduced the number of rabies cases in people from 250 to one in three years. And in Costa Rica, in league with the Ministry of Health, volunteers in elementary schools teach and demonstrate how to brush teeth, which kids do as a daily ritual with their teachers.

The majority of volunteers are high school and college students, although volunteers of all ages and nationalities are welcome to join. One of the organization's strong points is "allowing young people to run the show in the field." All of the trainers in the field are previous volunteers, so Amigos operates a youth-training program as well.

A prerequisite is being able to communicate in Spanish, because volunteers live in pairs or threes in remote villages and will need the language to survive. In training sessions before going into the field, volunteers learn Spanish, whatever health techniques they will be teaching, and the cultural habits of the host country. They also learn ways to ensure their own personal health while in the field.

OPERATION CROSSROADS AFRICA

Begun in 1958 by Dr. James H. Robinson, pastor of a Harlem church, after he traveled to Africa and realized how little the newly independent nations understood the United States, Crossroads was intended to foster friendship among young Africans and Americans. Using his experience with summer camps in Harlem, Dr. Robinson arranged for American students to work side by side with Africans toward a common goal. Today, college students spend two months in the summer building schools, working on farms, creating water supplies, driving a mobile health van to villages in the bush, or working in health clinics, where they give inoculations and teach sanitation. If the program sounds like the Peace Corps, it should: President John F. Kennedy created the Peace Corps after hearing about Crossroads, and Dr. Robinson served on the Peace Corps Committee for many years.

In December 1990, Crossroads will begin 3- and 6-month programs for college students to teach English in Namibia, whose leaders would like to establish English as the country's official language.

Fifty to 60 percent of participants receive college credit. Crossroads also runs a similar program in the Caribbean for high school students. Housing is usually in university or school dormitories or youth hostels. Groups live and work together with experienced project leaders.

The two-month summer program in Africa is $3,500; in the Caribbean, $2,500.

Chapter Eight

Practical Information

Health Tips

Before you head off to a little-known place in the world, let alone a rural district of a little-known place in the world, see your physician. Some physicians subscribe to a computer software service called TRAVAX, which gives a health-related readout of a country. Ask the middleman organizers what shots you will need; they, or their project leaders, are already familiar with the field and will know what specific things you can do to avoid illness. Be sure to allow enough time to get all the shots you will need, and make sure your previous inoculations are up to date.

The following are general tips offered by previous volunteers and project personnel.

AIDS: The virus is not airborne; it is very short-lived outside the body; and bleach kills it on surfaces, health workers say. It is transmitted sexually or through infected needles. Anywhere in the world, take precautions before having sex with someone who is not your sole partner. Use rubber condoms. It is inconvenient to carry pints of your own blood with you; but some travelers take along pre-packaged sterilized needles in the event they might have to give a blood sample. For up-to-date information on AIDS, call the Centers for Disease Control AIDS Information number: (404) 639-1610.

Altitude sickness: This can be very serious. If you have made a fast ascent to at least 9,000 feet, and you feel drowsy, have slurred speech, a crashing headache, and no appetite, immediately either breathe oxygen or go down 1,000 feet. Under no circumstances should you sleep. If you have a cough and are spitting blood, get to a medical facility immediately: these are symptoms of acute pulmonary edema. It is wise to make ascents slowly. If you will be working at high altitudes, it is better to descend to sleep at a lower altitude. Young people tend to be more susceptible to altitude sickness.

Blisters: Eliminating the source of friction is the best way to speed the cure of a blister. But if you have a blister on your foot and you can not change shoes, put a piece of moleskin between the blister and your shoe. Air helps to cure a blister; seawater is also a good cure.

Cleanliness: Most doctors say that having clean hands is probably the easiest and most important thing you can do to stay healthy. Moist towels in foil packages can keep hands clean.

Coconuts: If you figure out a way to get a coconut open, be careful when drinking one that is completely ripe, or red, because its milk is a laxative.

CPR: This is a life-saving skill that is good to have, but be sure to renew your certification each year.

Diarrhea: Seasoned travelers swear by Pepto-Bismol, which you can carry as tablets or in a bottle. Lomotil, which is a prescription drug in the United States, does not work for all kinds of bacterial infections that cause diarrhea and can in fact take longer to solve the problem. Drink as much water as you can to replace lost fluids; fruit juice or a simple solution of salt and sugar in water helps replace lost salts.

Drinking water: Few rural communities in less developed countries have potable water. Be careful when buying bottled water, however; it might not be any better. Hot tea or coffee is usually safe. Old hands swear by local beer, which is usually safe, or fruit juices and Coca-Cola, Pepsi Cola, and their derivatives, which are almost but not quite ubiquitous.

If you boil water for drinking, boil it at least for 5 minutes at sea level (some say boil for 10 minutes) and 15 minutes at high altitudes (because water reaches a boiling level faster in higher altitudes). You can also buy water-purifying pills in most outdoors-outfitting stores; or use six drops of iodine per quart of water and let stand for 30 minutes. Brush your teeth with bottled or boiled water.

Fish for dinner: If the fish is from fresh water, it is probably edible. But if it is from the ocean, it may be poisonous. Generally, fish without scales make dicey dinners. If it's poisonous, you'll know immediately because your mouth will sting. Within the next 30 hours, you might develop vomiting, diarrhea, paralysis, and convulsions. If you think you've eaten something poisonous, drink lots of warm saltwater to induce vomiting, and get medical help.

Fresh water: Avoid bathing in stagnant fresh water in the tropics, because of snail-borne larvae that penetrate the skin and cause bilharzia, or schistosomiasis, which infects the internal organs. If you have the option of bathing in the ocean, do so. Outdoors outfitting stores sell soap that works well in saltwater.

Frostbite: Red, gray, white, gangrene, in that order, is what to remember if you think you have frostbite. Warm the affected parts of your body slowly in warm water.

Hypothermia: To bring someone back from extreme chilling in which the core temperature drops and shivering stops, wrap the person in blankets, or, even better, share your own body heat.

Jet lag: Recent studies at Harvard University Medical School indicate that such a simple thing as exposure to light at your destination dramatically shortens jet lag. Jet lag is the feeling of total fatigue and irritability that results when the body's circadian rhythm is disrupted. The researchers recommend walking or sitting outside in natural light for as long as you can after you arrive.

Flying north and south will not produce noticeable jet lag; flying west will produce less jet lag then flying east. Some travelers suggest mild exercise after you arrive, such as taking a walk or going for a short run.

Malaria: The female Anopheles mosquito needs protein from blood to develop her eggs. In the process of taking your blood, she deposits the malaria protozoa, which will manifest itself in 6 to 12 days and cause high fever, chills, and intense sweating. Before you travel, ask your doctor what is the best current antimalarial medication for *you* to take. Each medication can have adverse side effects for some people. To make matters worse, the malaria protozoa developed an immunity to chloroquinine within the last decade at the same time that the tough Anopheles mosquito developed a resistance to insecticides.

According to the World Health Organization, the best defense against malaria is not to be bitten. Therefore, you should know certain things about Anopheles mosquitoes:

• They attack only between sunset and sunrise.

• In most countries, they live only in rural areas, not in cities—except in Africa.

• They breed anywhere around stagnant water, even in a puddle in a depression left by a footprint.

• They do not buzz or hover.

• They do not rest on walls with their bodies parallel to the wall: their rear ends stick up.

• They are attracted to your warmth and your exhaled breath.

The World Health Organization and IAMAT (please see below) offer the following recommendations:

• Wear long-sleeved shirts and long pants.

• Avoid dark colors; wear beige or yellow or light blue.

• Sleep in screened rooms.

• Sleep under a mosquito net that is tightly secured under your mattress.

- Burn mosquito coils.
- Spray the sleeping area with insect repellent.
- Cover the exposed parts of your body with insect repellent.

Malaria will not incubate in less than 6 days, so if you develop a fever less than 6 days after you arrive, do not assume it is malaria. Get to a doctor. Be sure to take your antimalarial medication regularly. One final note: some travelers claim that Avon's hand cream Sofskin, which was not developed as an insect repellent, is the best insect repellent around.

Medicine chest: Add some anti-itch lotion, along with Band-Aids, personal drugs, topical antibiotics like Neosporin or Bacitracin, antacids, aspirin, cold medicine, and anything else you think you need or that the project leaders advise you to bring. Some doctors think taking a little tetracycline each day will guard against diarrhea. But it will also make you photosensitive, and you could get a bad sunburn.

Salads: The old travelers' rule is: cook it, peel it, or forget it.

Scuba: Scuba diving increases the nitrogen in your blood, which means that you should carefully consider how quickly you dive after flying or fly after diving. The pressure in the cabin of an airplane flying at 36,000 feet is equal to the pressure on land at an altitude of about 7,000 feet. This means that you can access less oxygen in your body.

A spokesman for the Divers Alert Network, a private nonprofit research and safety organization for sports divers, recommends postponing flying for 24 hours after a single noncompression dive and 48 hours after a compression dive. These recommendations, the spokesman emphasizes, are stricter than the usual U.S. Navy guidelines "because the navy is on a mission; recreationals aren't."

If you are planning to dive soon after flying, the altitude of the lake or body of water should be taken into consideration. Lake Titicaca, for example, is more than 12,000 feet high. Consult your dive shop or call the Divers Alert Network for reference to conversion tables for the country in which you will be diving.

In the water, no matter what you will be doing, experienced divers say to plan your dive, and dive your plan; and use the buddy system.

The Divers Alert Network, or DAN, is based at Duke University in Durham, North Carolina. It runs a 24-hour, 365-day-a-year emergency hotline that provides advice from diver medical technicians (DMTs) and dive physicians, who, DAN says, are rare.

DAN also offers insurance that covers decompression sickness, arterial gas embolism, and pulmonary barotrauma, as well as recompression services and emergency evacuation. The annual individual membership for $15.00 includes a subscription to its newsletter and a copy of its underwater accident report; "prepared" membership for $40.00 includes the above, plus insurance. Their information number is (919) 684-2948 or (800) 446-2671. Address: Divers Alert Network, Box 3823, Duke University Medical Center, Durham, NC 27710.

Seasickness: Be sure to take medication *before* you go aboard. If you get caught without it, stare at the horizon and do some mental gymnastics such as counting backwards by 9s or declining French verbs. Munching on soda crackers keeps the stomach busy; but it is wise to avoid eating much more than that. If you know you are susceptible to sea sickness, take along a slow-release medication that comes in the form of a small, round disk that sticks to the skin behind the ear. Put it on at least two hours before going to sea; it lasts for three days.

Sleep: It is the quality of sleep that counts on a field trip. REM (rapid cyc movement, or dreaming) sleep, for example, seems to promote absorption of new memories and makes it easier to learn new tasks. Researchers have not yet been able to identify exactly what promotes good REM sleep, but drinking alcohol before going to bed interferes with it. Avoid caffeine after dinner; and if you can't sleep, go to bed later. Most sleeping pills will leave you drowsy the next day. Also, the antimalaria drug Mefloquinine can inhibit physical and mental coordination for a day after you take it. On work projects, most volunteers sleep very well. After work, you might have a chance for a 15-minute power nap.

Stress: Headphones are probably the best invention since Valium for stress. Psychologists advise investing in the kind of tapes that guide you into a relaxed state or that play nature sounds. Travel doctors emphasize that all travel tends to be stressful; travel to a work camp in the rural outback with a handful of total strangers is off the stress scale.

Sunburn: Use a sunscreen of at least SPF-15 and "full spectrum," which protects against UVA, UVB, and infrared rays. Make sure it is waterproof. Use a total block on your lips and keep reapplying it. If you get a sunburn, aloe vera gel is a good balm to restore your skin, and cover your skin or stay out of the sun for a while.

Sunglasses: Glare from mountain snow or off the ocean can cause a type of conjunctivitis that can result in watery eyes, difficulty in keeping your eyes open, and a feeling of grittiness. The best sunglasses block out 90 to 100 percent of UV sun rays. Make sure they are closed on the sides. Carry an extra pair with you.

Travel Tips

VISAS

Your middleman organization will tell you if you need a visa and might help you get one. Or you can call the local consulate, if there is one in your community. The *USA Today* 24-Hour Weather Hotline gives a wealth of information about visas, foreign currency rates, U.S. Department of State advisories about foreign travel, as well as the weather both here and abroad for 4- to 10-day predictions. That number is 1-900-370-USAT. Or you can contact the following visa service, which will process your visa in 7 to 21 days (depending on the country) for a modest fee: Visas International, 3169 Barbara Court, Los Angeles, CA 90068, (213) 850-1192.

INSURANCE

You can buy all kinds of travel insurance to cover everything from lost luggage to problems resulting from a hijacked airplane. Most short-term travel policies offer compensation for canceled flights as well as emergency health care. For scuba insurance, please see the Divers Alert Network, under Scuba, above.

For health insurance, look for a company that offers assistance in finding an English-speaking doctor, or a translator, and make sure you are covered for appropriate medical care, including coverage for evacuation and travel back to the United States. Find out what your own health insurance covers, then buy whatever you need to supplement that.

Middleman organizations make suggestions about insurance. The Foundation for Field Research offers California Workman's Compensation as part of the expedition price. This covers all volunteers working on any project with a California-based organization. FFR suggests buying supplementary trip cancellation insurance.

Below is a list of travel insurance companies that will send you a brochure:

Access America: 1-800-284-8300
HealthCare Abroad: 1-800-237-6615
International SOS: 1-800-523-8930
Travel Assistance International: 1-800-821-2828
Holidair Insurance Services Inc.: 1-800-525-2528

They all differ slightly; general coverage costs about $50 for 14 days.

HEALTH INFORMATION
Bon Voyage in Good Health is a video produced by health care specialists at the University of Minnesota's Boynton Health Service. It is available for about $20 from Shoreland Medical Marketing, 5827 West Washington Street, Milwaukee, WI 53208-1652.

The Centers for Disease Control in Atlanta provides up-to-date information on shots, malaria, AIDS, hepatitis, and other diseases. Call (404) 639-2572.

Health Information for International Travel, an annual publication of the Centers for Disease Control, is available for under $5 from your local government printing office bookstore; or contact Superintendent of Documents, U.S. Government Printing Office, Washington, D.C. 20402; (202) 783-3238.

IAMAT, International Association for Medical Assistance to Travelers, a nonprofit group, offers—for a donation—excellent, current information on world health hazards. It publishes regularly updated lists of English-speaking doctors in other countries; a worldwide immunization chart, with special recommendations for short- and long-term travel; complete information on malaria and antimalarial medication with a risk chart and protection guide; a risk chart of schistosomiasis and chaka disease; as well as world climate charts, with information on the safety of drinking water.

IAMAT was established by Toronto physician Vincenzo Marcolongo about twenty years ago when he saw the need for an international network of doctors for travelers. "When the relations of men are expressed in terms of cooperation and human endeavor," he wrote shortly before he died, "we become conscious and deeply aware that we belong to each other."

The information is available for a donation from IAMAT, 417 Center Street, Lewiston, NY 14092; (716) 754-4883.

International Travel and Health, Vaccination Requirements and Health Advice, published annually by the World Health Organization in Geneva, offers all that its title says. Your local library might have a copy, or contact: United Nations Publications, Sales Section, Room DC2-853, Dept. 701, New York, NY 10017, (212) 963-8302.

A company called Medic-Alert will engrave a bracelet or a necklace with a person's primary health condition and special problems, including allergies, along with a telephone number for Medic-Alert and the bearer's member number. That way, any attending physician can retrieve complete health information on a patient in a hurry.

For the cost listed below, you provide your pertinent medical data, and Medic-Alert will put you on file and send you a bracelet or a necklace with the following choice: $25 for stainless steel; $35 for sterling silver; $45 for gold plate. To update or change your file will cost $5. Contact Medic-Alert, P.O. Box 1009, Turlock, CA 95381; (800) 344-3226

SAFETY

Leaving a project in a small village in Peru one day, volunteer Robert Schilling, who goes on several projects a year, was standing on the street waiting for a bus when someone distracted him. "I should have known," Schilling said, realizing he had been set up. His bag containing his tickets, traveler's checks, money, passport, identification papers—everything—was gone. Left with a credit card and some local money, he went to Peru Airlines, where an agent told him to go to Lima. On the plane, he met a Peruvian and told him his story, and the man told him to fill out a police form but not to mention the village, because it could take years to resolve. "It didn't matter whether it happened in Lima or someplace else," Schilling said.

In Lima, he filled out the police form, then found he was in a double bind: he needed $40 to get a new passport and a passport to file the form to redeem his lost traveler's checks. Finally, he went to the American consulate where he scraped together enough local money to pay for his passport. The consulate also helped get him a ticket home.

"No matter where you are, someone's going to help you," says Schilling.

It pays to imagine the worst-case scenario and to prepare for it. This means carrying irreplaceable things close to your body in a money belt or inner pockets. Mark Richman of IRE cautions his volunteers to carry all camera equipment with them, because, in any country, suitcases are often rifled when they are checked.

If you must take two bags, carry one with you which contains a couple of days' clothes, maps, telephone numbers, drugstore items and prescription medicine, eyeglasses, and other necessities.

DRIVING

If you plan to drive overseas, check to see if your project allows it. Some do not. If you do drive, be aware of recent statistics from the World Health Organization which indicate that the leading cause of death among tourists is automobile accidents. Safety starts with the car: in some countries, auto rental companies will rent anything that is able to wobble out of the lot. At the very least, be sure your car has brakes.

If you drive overseas, you might need an international

driver's license. Not all countries require them; check first to see if your own driver's license will suffice. The AAA will provide this information and will service your international driver's license. You have to visit the office, bring two passport pictures, and your own driver's license. It costs under $10. You can also buy foreign maps there.

FILM AND VIDEO

Not all airport X-ray technology is the same; in some less developed countries, it can cook your film. Generally, the more times you go through security, the greater the risk to your film. Some photographers carry lead-lined bags; it is easier to keep your film and camera separate and hand them to the guards.

Video film is not affected by X rays, but it can be erased by metal detectors. Don't take it with you through the detector.

LEGAL PROBLEMS

Do not buy black market currency, and do not get involved with drugs. Penalties in some countries are unremittingly stiff, no matter who you are back in Boise. If, for any reason, you find yourself entangled in a difficult situation, ask for help at the nearest American consulate.

CLOTHES

Whether you shop at Woolworth's or in Patagonia, comfort is the key. If you will be working at night in a cold climate, thermal underwear is a necessity; otherwise, take a lot of warm clothes to layer. In the tropics, cotton and linen are cooler than synthetics or silk, especially if you have to keep covered up in a jungle.

Hats: The ozone layer being what it is, hats might be the most important thing you could bring. In Arctic-type climates, get a hat that covers your cheeks, too.

Jeans: Seasoned workers say that the most practical jeans are part polyester for one good reason: they dry faster than all-cotton jeans.

Rain gear: A waterproof jacket, cape, or coat that can fold up into a tiny pouch is ideal in rainy places. In the tropics, few people wear raincoats, because rain showers are quick, and raincoats are hot.

Shirts: Long-sleeved T-shirts are better than buttoned shirts in the jungle because nothing can fly in, and nothing can get caught on branches.

Shoes: Shoes are the most important clothing item to think about. Make sure you wear the right shoes for your activity, and be sure they are already comfortable before you go. The right shoes will make all the difference in climbing on rugged terrain, in walking through damp forests, in walking along coral beaches, in hiking, in trudging through snow, and in walking where a lot of snakes or scorpions live. Some people never take off their sneakers. A lot of people take flip-flops for showers.

Socks: Veteran hikers know that cotton socks can cause grief. If cotton socks get wet, they take a long time to dry, and they bunch up into weird shapes inside your shoes. Cotton or wool socks that are part acrylic dry faster and keep their shape.

Other things some travelers suggest taking:
a space blanket
extra flashlight batteries
a sweater
a drinking-water bottle
an alarm clock
a pocketknife
a language dictionary
work gloves
a fanny pack
whatever creature comforts (e.g., chocolate) make your
life nice.

Creatures to Avoid

There are not that many poisonous creatures on earth, and
usually they are shy in addition to being well defended.
Most encounters are rare; still, it pays to know if you're star-
ing into the eye of a poisonous or a nonpoisonous snake.
Find out in advance what creatures are likely to be in the
area you are traveling to. Do some library research, and
when you get there, talk to the locals. Following are a few
of the better-known venomous types to watch out for.

ARTHROPODS AND SNAKES
Some countries have more than others, and the tropics produce large ones—centipedes, scorpions, spiders, and snakes.

Scorpions: In a movie, Clint Eastwood once leveled a
scorpion with a well-placed ball of spit. Doctors don't recommend that, however, even if scorpions drown easily.
Insecticide is a better idea. Be sure to check for them wher-

ever it's warm and cozy—your shoes, boots, and bed. Not all are poisonous; the deadliest ones are in North Africa. Reactions to scorpion venom are rapid breathing, excessive salivation, and nausea. Get to a medical facility as soon as possible.

Spiders: The best-known poisonous spiders are the black widow, which has a tiny red dot on a shiny black body, and the brown recluse, which has a darker brown violin shape on its pale brown head. Both have bodies about half an inch long. Tarantulas have a lot of body hair, but people who keep them as pets say they are harmless. Spiders of all kinds like to gather at the latrine to eat insects. The trouble with spiders is that they are small and you can't tell when they might be testy. Everyone knows the names of the most famous deadly spiders, but no one knows how many deadly species there really are. Get to a doctor fast if you think you might have been bitten. Reaction to a black widow bite resembles acute appendicitis.

In Argentina, on a project studying the habits of an elusive animal that lived in the dense underbrush, a volunteer was bitten by a deadly spider. The project leader rushed him to the nearest hospital where doctors operated on his foot to remove the infection. Then for 10 days he stayed at a hotel nearby to go back for daily treatments. He became somewhat of a celebrity. "People from all over town would come to visit every day. They would bring me things and would talk about my foot."

Snakes: Half the battle with snakes is knowing something about them before you're in their territory. Ask the project leader what's out there, then read about them. Most snakes are harmless; a few are not; some are harmful and gigantic; some look harmless but are deadly. The fast and skinny, 14-foot-long African black mamba is thought to be one of the deadliest snakes in the world.

Preparation is the best defense. Wear boots and sleep off the ground, but bear in mind that some snakes sleep in trees. Make a lot of noise when you are walking where they might be. Pit vipers, which are always poisonous, have little slashes under their eyes. Snakes will not usually pursue a person. When they rear up, they can strike only to within one half the length of their body, so think and act quickly.

Outdoors outfitting stores sell snakebite kits; but if you have to use one, remember that you have about a half hour before the venom spreads to the rest of the body. Get to a medical facility as soon as you can: antivenim is effective only within 4 hours after the bite.

IN THE SEA
Scuba divers love to tell stories about encountering large, deadly creatures, but the small ones are more likely to be a problem.

Cone mollusks: Of all the things to worry about in the sea, this creature should be first on your list. All species of cones are deadly, yet they are likely to be the shells any casual collector will notice. Don't pick up the shells on a beach unless you are absolutely sure they are empty. The creature inside has a long stinger that shoots out like a harpoon and injects a poison that, in some cases, has caused death in minutes.

Coral: All coral abrasions sting; those from the fire coral (which is really a dull yellow), especially. The wound will not heal unless you rinse it with bleach or hydrogen peroxide or a baking soda solution. Then wash it and apply a topical antibiotic. Always wear shoes walking on or near coral.

Crown-of-thorns starfish: This pillow-size starfish has wavy, 12-inch quills that are fine and savagely sharp.

Spiny sea urchins have quills that can cause discomfort if you walk or fall on them. In both cases, getting the quills out can take longer than the discomfort from the sting.

Octopus: Only one species of octopus is dangerous: the blue banded octopus that lives on the reef off Indonesia and on the Great Barrier Reef.

Portuguese man-of-war: These are jellyfish with long feathery blue tendrils. They move through the water like ballet dancers and can cause a lot of grief. Even if you just touch parts of them that have washed up on the beach, you will feel the sting. Man-of-wars turn up in the Atlantic as well as the Pacific. Meat tenderizer sprinkled on the area of contact, after it is washed, will take away the pain. Some folks use vinegar.

Sea snakes: These black and white snakes are not large, but they are bright, swift, and unpredictable. The toxic effects begin within 20 minutes with malaise and anxiety, followed by muscle spasms (including difficulty breathing) and shock. Get out of the water and wrap a tourniquet between the wound and your heart until you get medical help.

Sea wasps: These small jellyfish have about a five-inch body and long tentacles and are as deadly as stonefish. Locals know if they are in the water; ask. Also ask locally if there exist any antidotes. They live in Australian waters.

Stonefish and lionfish: The deadliest of all poisonous fish, the stonefish resembles a rock and sits on the bottom in shallow water with its poisonous spines waving on its back like seaweed. Lionfish are easier to spot because they are very colorful, with stripes and feathery spines. In both cases, their sting can produce what observers describe as intense agony. Plunge the afflicted foot or leg into very hot water, if possible. This will help temporarily; and then get to a doctor as quickly as possible.

Check your local bookstores for guides to potentially dangerous creatures. A standard reference is *Venomous Animals and Their Venoms,* edited by Wolfgang Buchel and Eleanor E. Buckley (New York: Academic Press, 1971, 2 vols.). Useful for reef hazards off the Florida coast and in the Caribbean is Eugene H. Kaplan's *A Field Guide to Coral Reefs* (Boston: Houghton Mifflin, 1982).

SOME MEASUREMENTS AND USEFUL FIGURES

1 kilometer = 0.062 mile. To figure the approximate number of miles in kilometers, multiply by 0.6.

1 kilogram (or ''kilo'') = 2.2 pounds.

10 cm = 3.9 inches.

20 cm = about the span of your hand from the thumb to the little fingertip

1 meter = 39.3 inches

1 liter = slightly more than a quart

1 hectare = 2.5 acres
0° F = -18° C
30° F = -1° C
50° F = 10° C
70° F = 21° C
80° F = 27° C
90° F = 32° C
100° F = 38° C

"They say God protects idiots and tourists, so I guess I'm completely covered."

—Volunteer

Organizations Offering Environmental Opportunities

AFAR
Box 83
Andover, MA 01810
(508) 470-0840

American Hiking Society
1015 31st Street NW
Washington, D.C. 20007
(703) 385-3252
(Conducts projects for the upkeep of national parks)

Amigos de las Americas
5618 Star Lane
Houston, TX 77057
(800) 231-7796

Arcosanti
HC 74
Box 4136
Mayer, AZ 86333
(602) 632-7135

Caretta Research Project
Box A
Savannah Science Museum
4405 Paulsen Street
Savannah, GA 31405
(912) 355-6705

Caribbean Conservation Corporation
P.O. Box 2866
Gainesville FL 32602
(904) 373-6441
(Volunteers research and help save the endangered green turtle in Costa Rica)

CEDAM International
1 Fox Road
Croton-on-Hudson, NY 10520
(914) 271-5365

Center for American Archaeology
Box 366
Kampsville, IL 62053
(618) 653-4316

Project Ocean Search
The Cousteau Society
930 West 21st Street
Norfolk, VA 23517
(804) 627-1144

Crow Canyon Archaeological Center
23390 County Road K
Cortez, CO 81321
(800) 422-8975

Desert Tortoise Council
5319 Cerritos Avenue
Long Beach, CA 90805
(213) 422-6172

Earth Island Institute
300 Broadway
Suite 28
San Francisco, CA 94133
(415) 788-3666
(Uses volunteers on the Environmental Project in Central
America, the International Marine Mammal Project, and
others)

Earthwatch
P.O. Box 403
Watertown, MA 02272
(617) 926-8200

Elderhostel
80 Boylston Street
Suite 400
Boston, MA 02116
(617) 426-7788

Foundation for Field Research
787 S. Grade Road
P.O. Box 2010
Alpine, CA 92001
(619) 445-9264

Habitat for Humanity International
Habitat and Church Streets
Americus, GA 31709-3498
(912) 924-6935

Hawk Migration Association of North America
P.O. Box 3482
Rivermont Station
Lynchburg, VA 24503
(804) 847-7811

International Research Expeditions
140 University Drive
Menlo Park, CA 94025
(415) 323-4228

Intersea Research Inc.
P.O. Box 1667
Friday Harbor, WA 98205
(800) 346-3516

Massachusetts Audubon Society
Natural History Travel
South Grate Road
Lincoln, MA 01773
(800) 289-9504

National Audubon Society
950 Third Avenue
New York, NY 10022
(212) 832-3200

Smithsonian Research Expeditions Program
Smithsonian Institution
490 L'Enfant Plaza SW, Room 4210
Washington, D.C. 20560
(202) 287-3210

University Research Expeditions Program (UREP)
University of California
Berkeley, CA 94720
(415) 642-6586

WorldTeach
Harvard University
Cambridge, MA 02138
(617) 495-5527

Opportunities for Students

Operation Crossroads Africa, Inc.
475 Riverside Drive
New York, NY 10115
(212) 870-2210

The School for Field Studies
16 Broadway
Beverly, MA 01915-2096
(508) 927-7777
(Offers credit summer and semester programs for high
school and college students in field studies in conservation
biology in the United States and around the world. Projects
range from the ecology of bald eagles in Alaska to sustainable
development in Zimbabwe. Exams are given in the field;
credit is from Northeastern University. Costs cover tuition,
food, and lodging: $2,000 for summer; $8,000 for 3-month
semester program. Generous scholarship program; about
one-third of participants are supported by scholarship.)

School for International Training
(The Experiment in International Living)
P.O. Box 676
Kipling Road
Brattleboro, VT 05301-0676
(800) 451-4465
(The Experiment in International Living was the first
organization to develop the concept of homestay in foreign
countries in 1932, when Donald Watt, a professor at Syra-
cuse University, took students overseas to live with host
families. Today, college students spend a semester not only
living with host families but also doing ecological field
studies, such as environmental sustainability in Ecuador,
ecosystems along the banks of the Amazon, coastal studies
in Kenya, and programs in China, Madagascar, Tanzania,
and other Third World countries. Costs of $6,100 to
$9,100, depending on the country, include tuition, food,

210

lodging, airfare, and 16 college credits. A high school program focuses on shorter-term homestay and language learning.)

Student Conservation Association
P.O. Box 550
Charlestown, NH 03603
(603) 826-4301
(For high school students, SCA offers a month of living and labor in the wilderness, helping construct trails or bridges, or being engaged in a wildlife conservation effort. The only expenses are travel fares and personal camping equipment. For college students and above, SCA sponsors a Resource Assistant Program for 12 weeks year-round, in which participants act as assistants to national and state parks professionals.)

Deep Travel Organizations
The following are some organizations that offer opportunities for "deeper" travel into Third World countries and the Soviet Union, in which participants meet people on a grass-roots level and are able to interview and exchange ideas with political and social leaders in the country. Lodging is either with families or in modest hotels.

The Antaeus Group
David B. Sutton, Executive Director
P.O. Box 4050
Stanford, CA 94309
(415) 851-2977

(Their Expeditions Program sponsors socially and environmentally sensitive tours in which participants can get involved in local conservation projects.)

Central American Mission Partners (CAMP)
P.O. Box 10206
Oakland, CA 94610
(415) 451-0919
(Ecumenical organization sponsors trips to Central America to talk with political and social leaders and to become acquainted with the problems of life in Central America.)

Citizen Exchange Council
12 W. 31st Street
New York, NY 10001-4415
(212) 643-1985
(Organizes 2- and 3-week educational meet-the-people tours to the Soviet Union.)

Global Exchange
2141 Mission Street, #202
San Francisco, CA 94110
(415)255-7296
(A travel program called Reality Tours makes it possible for participants to talk with community leaders and meet people in the Third World in an effort to make helpful connections. Worldwide opportunities; cost: $300-$3,300 includes airfare, food, lodging, reading materials.)

Global Partners in Travel
717 E. Shaw Avenue
Fresno, CA 93710
(800) 735-8222
(Sponsors people-to-people travel in the Third World and the Soviet Union.)

Institute of Noetic Sciences
475 Gate Five Road, Suite 300
P.O. Box 909
Sausalito, CA 94966-0909
(800) 525-7985
(Founded by former astronaut Edgar Mitchell in 1973 "to expand knowledge of the nature and potentials of the mind and spirit and to apply that knowledge to advance health and well-being for humankind and our planet," the Institute sponsors informed, environmentally sensitive travel around the world, often linked to its interesting research projects, particularly mind-body healing, exceptional human abilities, and "emerging world views in science, business, and society.")

International Expeditions, Inc.
1776 Independence Court
Birmingham, AL 35216
(800) 633-4734
(An ecotour company)

One World Family Travel Network
c/o Lost Valley Center
81868 Lost Valley Lane
Dexter, OR 97431

(Publishes *Directory of Environmentally Responsible Travel* [$10.00] and *The Directory of Alternative Travel Resources* [$7.50].)

Pax World Foundation
4400 East-West Highway
Suite 130
Bethesda, MD 20814
(301) 657-2440
(Sponsors trips to various parts of the Third World to discuss issues with local leaders. Also, sponsors a program to plant trees in Antigua. Cost: $1,050 -$1,800, includes airfare from JFK, food, and lodging.)

Organizations Involved with Environmental Issues

American Anthropological Association
1703 New Hampshire Avenue, NW
Washington, D.C. 20009
(202) 232-8800
(Their annual publication, *Anthropology Field Schools,* lists various opportunities, mostly for students. $5 for members, $7 for nonmembers, plus an SASE with 56¢ postage.)

Archaeological Institute of America
675 Commonwealth Avenue
Boston, MA 02215
(617) 353-9361
(They publish *Archaeological Fieldwork Opportunities Bulletin* annually: $10.50 for members, $12.50 for nonmembers, includes postage.)

214

The Arctic to Amazonia Project
Box 73
Strafford, VT 05072
(802) 765-4337
(Sponsors conferences that bring together Native Americans whose environment is threatened by industry and industrial representatives to foster an open dialogue. Will sponsor a conference entitled "Seeds of Hope" in 1991 to discuss economics and resource management programs based on tribal rather than industrial models. The project's director, Erik van Lennep, says their office has excellent sources of information for people looking for a particular environmental area in which they may be of help.)

**The Coolidge Center
for Environmental Leadership**
1675 Massachusetts Avenue, Suite 4
Cambridge, MA 02138-9719
(617) 864-5085
(Publishes a calendar listing major worldwide conferences and lectures and films relating to international development and the environment. Also publishes a quarterly newsletter. Yearly membership: $25.)

Cultural Survival
11 Divinity Avenue
Cambridge, MA 02138
(617) 495-2562
(This is a nonprofit, human rights organization that works with indigenous people around the world, focusing on land rights and training in natural resource management. Their

quarterly magazine is a gold mine of information about abuses in the Third World. Call for membership information; they also need volunteers in their office.)

Conservation International
1015 18th Street NW, Suite 1000
Washington, D.C. 20036
(202) 429-5660
(This organization is involved with the restoration of rain forests.)

Co-Op America
2100 M Street NW, Suite 403
Washington, D.C. 20063
(800) 424-2667
(Their Travel Link Program will direct members to socially responsible alternative travel opportunities from listings in a directory they have compiled. Trips vary from 2 to 6 weeks and can be styled to special interests, from learning a foreign language in a host country to helping people in the Third World work on projects. Membership of $20 a year provides a quarterly magazine, a twice-a-year catalog linking consumers with environmentally sensitive products, and opportunities for investments and insurance for nontraditional and traditional health care.)

Defenders of Wildlife
1244 19th Street NW
Washington, D.C. 20036
(202) 659-9510
(Uses volunteers to protect animal habitats.)

Ducks Unlimited
One Waterfowl Way
Long Grove, IL 60047
(312) 438-4300
(Uses volunteers to help develop and maintain waterfowl habitats.)

Earth Team
Soil Conservation Service
South Bldg., Room 4237
Box 2890
Washington, D.C. 20013
(202) 382-0430
(Needs volunteers to work with farmers and ranchers to preserve soil against erosion and flooding.)

Friends of the Earth
218 D Street SE
Washington, D.C. 20036
(202) 544-2600 (Washington office)
(206) 633-1661 (Seattle office)
(An international advocacy group that deals in issues that directly affect our lives, such as toxic wastes. Uses volunteers in their offices in all capacities from lobbying to stuffing envelopes.)

Greenpeace
1436 U Street NW
Washington, D.C. 20009
(202) 462-1177
(Actively involved in preventing environmental abuses. Call for membership information.)

Izaak Walton League of America
1401 Wilson Boulevard
Level B
Arlington, VA 22205
(703) 528-1818
(Several volunteer programs help protect land, water, air resources.)

National Wildlife Federation
1400 16th Street NW
Washington, D.C. 20036-2266
(202) 797-6800
(Education association involved in conservation and wise use of natural resources.)

The Nature Conservancy
1815 N. Lynn Street
Arlington, VA 22209
(703) 841-5300
(Uses volunteers in protection of wildlife in sanctuaries.)

North American Wildlife Foundation
102 Wilmot Road
Suite 410
Deerfield, IL 60015
(708) 940-7776
(Involved with preservation of North American wildlife.)

The Oceanic Society Expeditions
Bldg. E
Fort Mason Center
San Francisco, CA 94123
(800) 326-7491
(Promotes conservation of marine life; sponsors naturalist tours.)

Pacific Whale Foundation
101 N. Kihei Road
Kihei, HI 96753
(808) 879-8860; (800) 942-5311

The Peregrine Fund, Inc.
5666 West Flying Hawk Lane
Boise, ID 83709
(208) 362-3716
(Works to recover falcons throughout the world.)

Sea Shepherds
Conservation Society
Box 7000-S
Redondo Beach, CA 90277
(213) 373-6979
(Actively stops mass animal slaughter, such as seal hunts, dolphins in tuna nets, etc. Really needs volunteers, especially any with marine skills. Sponsors two expeditions a year. Cautions that there is danger involved.)

The Sierra Club
730 Polk Street
San Francisco, CA 94109
(415) 776-2211
(Call for membership information. The Sierra Club sponsors ecologically sensitive outings, and local chapters run volunteer wilderness outings programs for inner-city kids and the handicapped.)

SOBEK
P.O. Box 1089
Angels Camp, CA 95222
(800) 777-7939
(An ecotourism company.)

Western Foundation for Raptor Conservation
P.O. Box 35706
1420 Carlisle, #202
Albuquerque, NM 87176
(505) 255-7622
(Uses volunteers in banding program.)

Wildlife Information Center
629 Green Street
Allentown, PA 18102
(215) 434-1637
(Uses volunteers in hawk-counting and other projects.)

World Wildlife Fund
1250 24th Street NW
Washington, D.C. 20037
(202) 293-4800

(Sponsors programs to protect African elephants and rain forests and to study resource use. Call for membership information.)

Zero Population Growth
1400 Sixteenth Street NW
Suite 320
Washington, D.C. 20036
(202) 322-2200
(Uses volunteers for various projects involved with population concerns.)

Magazines
The following magazines often list opportunities for environmental travel and volunteer involvement or publish articles on environmental issues: *Buzzworm, E, New Age Journal, Greenpeace, Outside, Mother Earth News, World Watch, Mother Jones, Archaeology, Smithsonian, Garbage.*

Books
Two books, published by the Council on International Educational Exchange (CIEE), are helpful:
Volunteer: Guidebook to Volunteer Services in the U.S. and Overseas (NY: CIEE, 1990) lists short-, medium-, and long-term volunteer opportunities. $6.95, plus postage: $1.00 book rate; $2.50 first class.
Work, Study, and Travel Abroad: The Whole World Handbook (New York: St. Martin's Press, published biannually). $10.95, plus postage: $1.00 book rate; $2.50 first class.
  Order from: Cheryl Jeffries, CIEE Pubs. Dept., CIEE, 205 E. 42nd Street, New York, NY 10017; (212) 661-1414.

Whale breaching off the Alaskan coast. Photo: Cindy D'Vincent/INTERSEA

Endangered Species

The Office of Endangered Species and the U.S. Fish and Wildlife Service annually publish *Endangered and Threatened Wildlife and Plants*. A copy is available at no charge from: Publications Unit, U.S. Fish and Wildlife Service, Washington, D.C. 20240; (202) 358-1711.

The International Council for Bird Preservation (ICBP) published *The Endangered Birds of the World, The ICBP Bird Red Data Book*, compiled by Warren B. King (Washington, D.C.: Smithsonian Institution Press, 1981). The ICBP and IUCN published *Threatened Birds of Africa and Related Islands* (Cambridge, England, 1985). *Birds to Watch* is a checklist of threatened birds in the world, intended as a companion to the Red Data Books. It is available for British £12.05 from: ICBP, 32 Cambridge Road, Girton, Cambridge CB30PJ, United Kingdom.

The International Union for Conservation of Nature and Natural Resources (IUCN) has compiled Red Data Books on endangered invertebrates (Gland, Switzerland, 1983); plants (Morges, Switzerland, 1978); amphibia and reptilia (Gland, Switzerland, 1982); freshwater fish (Gland, Switzerland, 1977); the primates of Africa (Gland, Switzerland, 1988); and animals and plants of the Soviet Union (Gland, Switzerland, 1978).

The following is a list of endangered mammals in the Western Hemisphere, as compiled by Jane Thornback and Martin Jenkins for the *IUCN Mammal Red Data Book* (Gland, Switzerland, 1982). The *Australasian* region includes Australia, New Zealand, and the island of New Guinea. The *Nearctic* region covers all of North America, including Greenland, the Aleutian Islands, and northern Mexico. The *Neotropical* region includes southern Mexico, all Caribbean islands, and Central and South America. 'Ex' denotes extinct in that country and '?' denotes the lack of confirmation of a mammal's presence.

Australasian Region

AUSTRALIA
Toolache Wallaby (Ex)
Bridled Nailtail Wallaby
Crescent Nailtail Wallaby (Ex)
Central Hare-wallaby (Ex)
Rufous Hare-wallaby
Eastern Hare-wallaby (Ex)
Banded Hare-wallaby
Proserpine Rock-wallaby
Desert Rat-kangaroo
Brush-tailed Bettong
Burrowing Bettong
Broad-faced Potoroo (Ex)
Long-footed Potoroo
Leadbeater's Possum
Northern Hairy-nosed
 Wombat
Pig-footed Bandicoot (Ex)

Western Barred Bandicoot
Desert Bandicoot (Ex)
Greater Bilby
Lesser Bilby (Ex)
Dibbler
Red-tailed Phascogale
Long-tailed Dunnart
Sandhill Dunnart
Julia Creek Dunnart
Numbat
Thylacine (Ex)
Ghost Bat
Rabbit-eared Tree-rat (Ex)
Lesser Stick-nest Rat (Ex)
Great Stick-nest Rat
Short-tailed Hopping-mouse
 (Ex)
Northern Hopping-mouse

Long-tailed Hopping-mouse
(Ex)
Big-eared Hopping-mouse
(Ex)
Darling Downs Hopping-
mouse (Ex)
Alice Springs Mouse (Ex)
Shark Bay Mouse
Hastings River Mouse
False Water-rat
Central Rock-rat
Dugong

INDONESIA (IRIAN JAYA)
Long-beaked Echidna
Doria's Tree-kangaroo
Stein's Cuscus
Black-spotted Cuscus
Clara Bandicoot
Dugong

PAPUA NEW GUINEA
Long-beaked Echidna
Doria's Tree-kangaroo
Goodfellow's Tree-kangaroo
Black Dorcopsis Wallaby
Papuan Dorcopsis
Stein's Cuscus
Woodlark Island Cuscus
Black-spotted Cuscus
Clara Bandicoot
Dugong

Nearctic Region
CANADA
Vancouver Island Marmot
Wolf
Polar Bear
Black-footed Ferret (Ex)

Eastern Cougar

DENMARK (GREENLAND)
Wolf
Polar Bear

MEXICO
Guatemalan Howler Monkey
Geoffroy's Spider Monkey
Volcano Rabbit
Mexican Grizzly Bear
Ocelot
Jaguarundi
Margay
Jaguar
Guadalupe Fur Seal
Caribbean Manatee
Central American Tapir
Cedros Island Mule Deer
Lower California Pronghorn
Sonoran Pronghorn

U.S.A.
Indiana Bat
Gray Bat
Hawaiian Hoary Bat
Ozark Big-eared Bat
Virginia Big-eared Bat
Delmarva Fox Squirrel
Utah Prairie Dog
Morro Bay Kangaroo Rat
Texas Kangaroo Rat
Salt-marsh Harvest Mouse
Beach Vole
Silver Rice Rat
Wolf
Red Wolf
Mexican Grizzly Bear
Polar Bear

Black-footed Ferret
Ocelot
Eastern Cougar
Florida Cougar
Jaguarundi
Jaguar
Hawaiian Monk Seal
Caribbean Manatee
Tule Elk
Key Deer
Columbian White-tailed Deer
Sonoran Pronghorn

Neotropical Region

ARGENTINA
Black Spider Monkey (?)
Giant Anteater
Giant Armadillo
Lesser Pichi Ciego
Greater Pichi Ciego
Short-tailed Chinchilla (?)
Maned Wolf
Bush Dog
Spectacled Bear (?)
La Plata Otter
Southern River Otter
Giant Brazilian Otter
Ocelot
Andean Cat
Little Spotted Cat
Jaguarundi
Margay
Jaguar
Chacoan Peccary
Vicuna
South Andean Huemul
North Andean Huemul
Marsh Deer
Argentinian Pampas Deer

BAHAMAS
Bahamian Hutia
Caribbean Manatee

BELIZE
Guatemalan Howler Monkey
Geoffroy's Spider Monkey
Giant Anteater
Ocelot
Jaguarundi
Margay
Jaguar
Caribbean Manatee
Central American Tapir

BOLIVIA
Emperor Tamarin
Goeldi's Marmoset
Brown Howler Monkey
Woolly Monkey
Black Spider Monkey
Giant Anteater
Giant Armadillo
Greater Pichi Ciego (?)
Short-tailed Chinchilla (?)
Bush Dog
La Plata Otter (?)
Giant Brazilian Otter
Spectacled Bear
Ocelot
Jaguarundi
Margay
Andean Cat
Jaguar
Chacoan Peccary
Vicuna
North Andean Huemul
Marsh Deer

BRAZIL
Buffy-headed Marmoset
White Marmoset
Buffy-tufted-ear Marmoset
Tassel-eared Marmoset
Bare-face Tamarin
Emperor Tamarin
Golden Lion Tamarin
Golden-headed Lion Tamarin
Golden-rumped Lion Tamarin
Goeldi's Marmoset
Masked Titi
White-nosed Saki
Southern Bearded Saki
Uakari (Red and White)
Black-headed Uakari
Brown Howler Monkey
Woolly Monkey
Woolly Spider Monkey
Long-haired Spider Monkey
Black Spider Monkey
Giant Anteater
Maned Sloth
Giant Armadillo
Brazilian Three-banded
  Armadillo
Thin-spined Porcupine
Small-eared Dog
Maned Wolf
Bush Dog
Spectacled Bear (?)
La Plata Otter
Giant Brazilian Otter
Ocelot
Little Spotted Cat
Jaguarundi
Margay
Jaguar
Caribbean Manatee

Amazonian Manatee
Marsh Deer

CHILE
Long-tailed Chinchilla
Short-tailed Chinchilla
Marine Otter
Southern River Otter
Andean Cat
Juan Fernandez Fur Seal
Vicuna
South Andean Huemul
North Andean Huemul

COLUMBIA
Cotton-top Tamarin
White-footed Tamarin
Goeldi's Marmoset
Uakari (Red) (?)
Black-headed Uakari
Woolly Monkey
Long-haired Spider Monkey
Brown-headed Spider Monkey
Geoffroy's Spider Monkey (?)
Giant Anteater
Giant Armadillo
Small-eared Dog
Bush Dog
Spectacled Bear
Giant Brazilian Otter
Ocelot
Little Spotted Cat
Jaguarundi
Margay
Jaguar
Caribbean Manatee
Amazonian Manatee
Mountain Tapir
Central American Tapir

Northern Pudu

COSTA RICA
Central American Squirrel
  Monkey
Geoffroy's Spider Monkey
Giant Anteater
Ocelot
Little Spotted Cat
Jaguarundi
Margay
Jaguar
Caribbean Manatee
Central American Tapir

CUBA
Cuban Solenodon
Cabrera's Hutia
Garrido's Hutia
Bushy-tailed Hutia
Dwarf Hutia
Little Earth Hutia
Caribbean Manatee

DOMINICAN REPUBLIC
Haitian Solenodon
Hispaniolan Hutia
Caribbean Manatee

ECUADOR
Goeldi's Marmoset (?)
Woolly Monkey
Long-haired Spider Monkey
Brown-headed Spider Monkey
Giant Anteater (?)
Giant Armadillo (?)
Small-eared Dog
Bush Dog
Spectacled Bear

Giant Brazilian Otter
Ocelot
Little Spotted Cat
Jaguarundi
Margay
Jaguar
Galápagos Fur Seal
Amazonian Manatee (?)
Mountain Tapir
Central American Tapir (?)
Northern Pudu

EL SALVADOR
Geoffroy's Spider Monkey
Giant Anteater (Ex)
Ocelot
Jaguarundi
Margay
Jaguar (Ex)
Central American Tapir (Ex)

FRENCH GUIANA
Black Spider Monkey
Giant Anteater
Giant Armadillo
Bush Dog
Giant Brazilian Otter
Ocelot
Little Spotted Cat
Jaguarundi
Margay
Jaguar

GUATEMALA
Guatemalan Howler Monkey
Geoffroy's Spider Monkey
Giant Anteater
Ocelot
Jaguarundi

Margay
Jaguar
Caribbean Manatee
Central American Tapir

GUYANA
Black Spider Monkey
Giant Anteater
Giant Armadillo
Bush Dog
Giant Brazilian Otter
Ocelot
Little Spotted Cat
Jaguarundi
Margay
Jaguar
Caribbean Manatee
Amazonian Manatee

HAITI
Haitian Solenodon
Hispaniolan Hutia
Caribbean Manatee

HONDURAS
Geoffroy's Spider Monkey
Giant Anteater
Ocelot
Jaguarundi
Margay
Jaguar
Caribbean Manatee
Central American Tapir

JAMAICA
Jamaican Hutia
Caribbean Manatee

NICARAGUA
Geoffroy's Spider Monkey
Giant Anteater
Ocelot
Little Spotted Cat (?)
Jaguarundi
Margay
Caribbean Manatee
Central American Tapir

PANAMA
Central American Squirrel
   Monkey
Brown-headed Spider Monkey
Geoffroy's Spider Monkey
Giant Anteater
Spectacled Bear (?)
Bush Dog
Ocelot
Little Spotted Cat
Jaguarundi
Margay
Jaguar
Caribbean Manatee
Central American Tapir

PARAGUAY
Giant Anteater
Giant Armadillo
Greater Pichi Ciego (?)
Maned Wolf
Bush Dog
La Plata Otter
Giant Brazilian Otter
Ocelot
Little Spotted Cat
Jaguarundi
Margay
Jaguar

Chacoan Peccary
Marsh Deer

PERU
Emperor Tamarin .
Goeldi's Marmoset
Red Uakari
Yellow-tailed Woolly Monkey
Woolly Monkey
Long-haired Spider Monkey
Black Spider Monkey
Giant Anteater
Giant Armadillo
Short-tailed Chinchilla (Ex)
Small-eared Dog
Maned Wolf
Bush Dog
Spectacled Bear
Marine Otter
Giant Brazilian Otter
Ocelot
Andean Cat
Little Spotted Cat (?)
Margay
Jaguar
Amazonian Manatee
Mountain Tapir
Vicuna
North Andean Huemul
Marsh Deer
Northern Pudu

SURINAME
Black Spider Monkey
Giant Anteater
Giant Armadillo
Bush Dog
Giant Brazilian Otter
Ocelot

Little Spotted Cat
Jaguarundi
Margay
Jaguar
Caribbean Manatee

TRINIDAD AND TOBAGO
Caribbean Manatee

URUGUAY
Maned Wolf (Ex)
La Plata Otter
Giant Brazilian Otter
Margay
Jaguar (Ex)
Marsh Deer (?)

VENEZUELA
Black-headed Uakari
Long-haired Spider Monkey
Black Spider Monkey
Giant Anteater
Giant Armadillo
Small-eared Dog (?)
Bush Dog
Spectacled Bear
Giant Brazilian Otter
Ocelot
Little Spotted Cat
Jaguarundi
Margay
Jaguar
Caribbean Manatee
Mountain Tapir

Oceans and Oceanic Islands

EAST PACIFIC OCEAN
Galápagos Fur Seal
Juan Fernandez Fur Seal
Guadalupe Fur Seal

NORTH PACIFIC OCEAN
Hawaiian Monk Seal
Dugong

SOUTH PACIFIC OCEAN
Dugong

CARIBBEAN SEA
Caribbean Monk Seal
Caribbean Manatee

ATLANTIC OCEAN
Caribbean Manatee

Index

Safety, 37, 40, 194-195
Sardinia, 105-107
Savannah, Georgia, 140-141
Schistosomiasis, 186, 193
Scholarships, 22, 49, 53, 60, 61, 170-171
Scientists, 77-88
Scuba, 42, 188-189
Sea projects, 151-163
Seasickness, 190
Senegal, 52
SHARE (UREP), 51, 125-129
Sharecroppers, 128-129
Singles, 41
Sleep, 36, 190
Sleeping quarters, 34-35
Smithsonian Research Expeditions Program, 47, 60-61; example of expedition, 129-133, 209
Snakes, 149, 198, 199-200
Social life, 41
Soleri, Paolo, 164, 167
South Pacific, 11-16, 155, 157
Spiders, 199
Stress, 190
Student opportunities, 177-182, 209-211
Sunburn, 191
Sunglasses, 191

Tanzania, 52, 125, 169-170
Tax deductibility, 23
Teaching opportunities, 169-172, 182
Thailand, 62
Ting, Irwin, 117-119
Tonga, Kingdom of, 11-16
Tourism, 1-9
TRAVAX, 183
Travel to the site, 33, 60, 96-99
Tunisia, 105
Turtles, 137-141; leatherback sea turtles, 137-140; loggerhead sea turtles, 140-141

University of California Research Expeditions Program (UREP), 19, 21, 22, 33, 42, 47, 50-54, 92, 106; examples of expeditions, 110-119;

SHARE projects, 123-129; teacher opportunities, 169- 171; 209

Virgin Islands: St. Croix, 138-140; St. John, 117-119
Video film, safety, 196
Visas, 191
Volcano Arenal, 130-133
Volcanoes, 129-133
Volunteer science, 5-6, 16-19, 21-26

Water: bathing, 186; drinking, 35, 185, 193
Webster, Grady, 115-117
Weekend projects, 59
Whales, 84-85; humpbacks, in Hawaii, 157-162; humpbacks, in Alaska, 162-163
Wolf howling, 136
Wolves, 134-136
World Health Organization, 194, 195
WorldTeach, 172, 209

X-rays in airports, 196

Other Books from John Muir Publications

Adventure Vacations: From Trekking in New Guinea to Swimming in Siberia, Richard Bangs (65-76-9) 256 pp. $17.95

Asia Through the Back Door, 3rd ed., Rick Steves and John Gottberg (65-48-3) 326 pp. $15.95

Being a Father: Family, Work, and Self, *Mothering* Magazine (65-69-6) 176 pp. $12.95

Buddhist America: Centers, Retreats, Practices, Don Morreale (28-94-X) 400 pp. $12.95

Bus Touring: Charter Vacations, U.S.A., Stuart Warren with Douglas Bloch (28-95-8) 168 pp. $9.95

California Public Gardens: A Visitor's Guide, Eric Sigg (65-56-4) 304 pp. $15.95 (Available 3/91)

Catholic America: Self-Renewal Centers and Retreats, Patricia Christian-Meyer (65-20-3) 325 pp. $13.95

Complete Guide to Bed & Breakfasts, Inns & Guesthouses, Pamela Lanier (65-43-2) 520 pp. $15.95

Costa Rica: A Natural Destination, Ree Strange Sheck (65-51-3) 280 pp. $15.95

Elderhostels: The Students' Choice, Mildred Hyman (65-28-9) 224 pp. $12.95

Environmental Vacations: Volunteer Projects to Save the Planet, Stephanie Ocko (65-78-5) 240 pp. $15.95

Europe 101: History & Art for the Traveler, 4th ed., Rick Steves and Gene Openshaw (65-79-3) 372 pp. $15.95

Europe Through the Back Door, 9th ed., Rick Steves (65-42-4) 432 pp. $16.95

Floating Vacations: River, Lake, and Ocean Adventures, Michael White (65-32-7) 256 pp. $17.95

Gypsying After 40: A Guide to Adventure and Self-Discovery, Bob Harris (28-71-0) 264 pp. $12.95

The Heart of Jerusalem, Arlynn Nellhaus (28-79-6) 336 pp. $12.95

Indian America: A Traveler's Companion, Eagle/Walking Turtle (65-29-7) 424 pp. $16.95

The Indian Way: Learning to Communicate with Mother Earth, Gary McLain (Young Readers, 8 yrs. +) (65-73-4) 114 pp. $9.95

The Kids' Environment Book: What's Awry and Why, Anne Pedersen (10 yrs. +) (65-74-2) 192 pp. $12.95 (Available 1/91)

Mona Winks: Self-Guided Tours of Europe's Top Museums, Rick Steves and Gene Openshaw (28-85-0) 456 pp. $14.95

The On and Off the Road Cookbook, Carl Franz (28-27-3) 272 pp. $8.50

Paintbrushes and Pistols: How the Taos Artists Sold the West, Schwarz and Taggett (65-65-3) 280 pp. $17.95

The People's Guide to Mexico, Carl Franz (65-60-2) 608 pp. $17.95

The People's Guide to RV Camping in Mexico, Carl Franz with Steve Rogers (28-91-5) 320 pp. $13.95

Preconception: A Woman's Guide to Preparing for Pregnancy and Parenthood, Brenda E. Aikey-Keller (65-44-0) 232 pp. $14.95

Rads, Ergs, and Cheeseburgers: The Kids' Guide to Energy and the Environment, Bill Yanda (8 yrs. +) (65-75-0) 108 pp. $12.95 (Available 2/91)

Ranch Vacations: The Complete Guide to Guest and Resort, Fly-Fishing, and Cross-Country Skiing Ranches, Eugene Kilgore (65-30-0) 392 pp. $18.95

Schooling at Home: Parents, Kids, and Learning, *Mothering* Magazine (65-52-1) 264 pp. $14.95

The Shopper's Guide to Art and Crafts in the Hawaiian Islands, Arnold Schuchter (65-61-0) 272 pp. $13.95

The Shopper's Guide to Mexico, Steve Rogers and Tina Rosa (28-90-7) 224 pp. $9.95

Ski Tech's Guide to Equipment, Skiwear, and Accessories, edited by Bill Tanler (65-45-9) 144 pp. $11.95

Ski Tech's Guide to Maintenance and Repair, edited by Bill Tanler (65-46-7) 144 pp. $11.95

Teens: A Fresh Look, *Mothering* Magazine (65-54-8) 240 pp. $14.95

A Traveler's Guide to Asian Culture, Kevin Chambers (65-14-9) 224 pp. $13.95

Traveler's Guide to Healing Centers and Retreats in North America, Martine Rudee and Jonathan Blease (65-15-7) 240 pp. $11.95

Understanding Europeans, Stuart Miller (65-77-7) 272 pp. $14.95

Undiscovered Islands of the Caribbean, Burl Willes (65-55-6) 232 pp. $14.95

Undiscovered Islands of the Mediterranean, Linda Lancione Moyer and Burl Willes (65-53-X) 232 pp. $14.95

A Viewer's Guide to Art: A Glossary of Gods, People, and Creatures, Shaw and Warren (65-66-1) 152 pp. $9.95 (Available 3/91)

22 Days Series
These pocket-size itineraries (4½″ x 8″) are a refreshing departure from ordinary guidebooks. Each offers 22 flexible daily itineraries that can be used to get the most out of vacations of any length. Included are not only ''must see'' attractions but also

little-known villages and hidden "jewels" as well as valuable general information.

22 Days Around the World, Roger Rapoport and Burl Willes (65-31-9) 200 pp. $9.95

22 Days Around the Great Lakes, Arnold Schuchter (65-62-9) 176 pp. $9.95 (Available 1/91)

22 Days in Alaska, Pamela Lanier (28-68-0) 128 pp. $7.95

22 Days in the American Southwest, 2nd ed., Richard Harris (28-88-5) 176 pp. $9.95

22 Days in Asia, Roger Rapoport and Burl Willes (65-17-3) 136 pp. $7.95

22 Days in Australia, 3rd ed., John Gottberg (65-40-8) 148 pp. $7.95

22 Days in California, 2nd ed., Roger Rapoport (65-64-5) 176 pp. $9.95

22 Days in China, Gaylon Duke and Zenia Victor (28-72-9) 144 pp. $7.95

22 Days in Europe, 5th ed., Rick Steves (65-63-7) 192 pp. $9.95

22 Days in Florida, Richard Harris (65-27-0) 136 pp. $7.95

22 Days in France, Rick Steves (65-07-6) 154 pp. $7.95

22 Days in Germany, Austria & Switzerland, 3rd ed., Rick Steves (65-39-4) 136 pp. $7.95

22 Days in Great Britain, 3rd ed., Rick Steves (65-38-6) 144 pp. $7.95

22 Days in Hawaii, 2nd ed., Arnold Schuchter (65-50-5) 144 pp. $7.95

22 Days in India, Anurag Mathur (28-87-7) 136 pp. $7.95

22 Days in Japan, David Old (28-73-7) 136 pp. $7.95

22 Days in Mexico, 2nd ed., Steve Rogers and Tina Rosa (65-41-6) 128 pp. $7.95

22 Days in New England, Anne Wright (28-96-6) 128 pp. $7.95

22 Days in New Zealand, Arnold Schuchter (28-86-9) 136 pp. $7.95

22 Days in Norway, Denmark & Sweden, Rick Steves (28-83-4) 136 pp. $7.95

22 Days in the Pacific Northwest, Richard Harris (28-97-4) 136 pp. $7.95

22 Days in the Rockies, Roger Rapoport (65-68-8) 176 pp. $9.95

22 Days in Spain & Portugal, 3rd ed., Rick Steves (65-06-8) 136 pp. $7.95

22 Days in Texas, Richard Harris (65-47-5) 176 pp. $9.95 (Available 11/90)

22 Days in Thailand, Derk Richardson (65-57-2) 176 pp. $9.95

22 Days in the West Indies, Cyndy & Sam Morreale (28-74-5) 136 pp. $7.95

"Kidding Around" Travel Guides for Children
Written for kids eight years of age and older. Generously illustrated in two colors with imaginative characters and images. An adventure to read and a treasure to keep.

Kidding Around Atlanta, Anne Pedersen (65-35-1) 64 pp. $9.95

Kidding Around Boston, Helen Byers (65-36-X) 64 pp. $9.95

Kidding Around Chicago, Lauren Davis (65-70-X) 64 pp. $9.95

Kidding Around the Hawaiian Islands, Sarah Lovett (65-37-8) 64 pp. $9.95

Kidding Around London, Sarah Lovett (65-24-6) 64 pp. $9.95
Kidding Around Los Angeles, Judy Cash (65-34-3) 64 pp. $9.95
Kidding Around the National Parks of the Southwest, Sarah Lovett 108 pp. $12.95
Kidding Around New York City, Sarah Lovett (65-33-5) 64 pp. $9.95
Kidding Around Philadelphia, Rebecca Clay (65-71-8) 64 pp. $9.95
Kidding Around San Francisco, Rosemary Zibart (65-23-8) 64 pp. $9.95
Kidding Around Washington, D.C., Anne Pedersen (65-25-4) 64 pp. $9.95

Automotive Repair Manuals

How to Keep Your VW Alive (65-80-7) 440 pp. $19.95
How to Keep Your Subaru Alive (65-11-4) 480 pp. $19.95
How to Keep Your Toyota Pickup Alive (28-81-3) 392 pp. $19.95
How to Keep Your Datsun/ Nissan Alive (28-65-6) 544 pp. $19.95

Other Automotive Books

The Greaseless Guide to Car Care Confidence: Take the Terror Out of Talking to Your Mechanic, Mary Jackson (65-19-X) 224 pp. $14.95
Off-Road Emergency Repair & Survival, James Ristow (65-26-2) 160 pp. $9.95

Ordering Information

If you cannot find our books in your local bookstore, you can order directly from us. Please check the "Available" date above. If you send us money for a book not yet available, we will hold your money until we can ship you the book. Your books will be sent to you via UPS (for U.S. destinations). UPS will not deliver to a P.O. Box; please give us a street address. Include $2.75 for the first item ordered and $.50 for each additional item to cover shipping and handling costs. For airmail within the U.S., enclose $4.00. All foreign orders will be shipped surface rate; please enclose $3.00 for the first item and $1.00 for each additional item. Please inquire about foreign airmail rates.

Method of Payment

Your order may be paid by check, money order, or credit card. We cannot be responsible for cash sent through the mail. All payments must be made in U.S. dollars drawn on a U.S. bank. Canadian postal money orders in U.S. dollars are acceptable. For VISA, MasterCard, or American Express orders, include your card number, expiration date, and your signature, or call (800) 888-7504. Books ordered on American Express cards can be shipped only to the billing address of the cardholder. Sorry, no C.O.D.'s. Residents of sunny New Mexico, add 5.625% tax to the total.

Address all orders and inquiries to:

John Muir Publications
P.O. Box 613
Santa Fe, NM 87504
(505) 982-4078
(800) 888-7504